NIGHTS
of
ICE

Also by Spike Walker

Working on the Edge

Nights of Ice

Spike Walker

St. Martin's Press

New York

≈

The photo of the author on the back flap was taken on the F/V *Alaska Trojan,* while fishing for halibut 120 miles off Kodiak Island.

Text design by Julie Jaycox.

Library of Congress Cataloging-in-Publication Data

Walker, Spike.
    Nights of ice / Spike Walker.—1st ed.
      p. cm.
    ISBN 0-312-15611-1
    1. Shipwrecks—Alaska.   2. Survival after airplane accidents, shipwrecks, etc. I. Title.
    G525.W238 1997
    979.8'051—dc21   97-6207
                                                    CIP

First Edition: October 1997

10 9 8 7 6 5 4 3 2 1

This book is for my mother,
Lorna McPherson Walker.

Contents

Acknowledgments

I would like to express my deep sense of personal gratitude to the scores of people who have contributed, in one form or another, to the creation of this book.

Foremost, I wish to thank Lynette Bishop for her encouragement and graciousness; and John M.G. Graham for his unwavering friendship and support.

I would also like to thank John Winther and Bart Eaton for providing several summers of work aboard the one-hundred-foot-long salmon packer *Theresa Marie,* and Conrad Johnson, skipper of the notorious ninety-foot *Nautilus.* I will always remember the long, invigorating months of work and camaraderie amid Alaska's awesome beauty as endless stretches of her pristine wilderness passed by.

The logistics involved in packing five million pounds of salmon from far out on the fishing grounds in southeast Alaska to the Trident Seafoods cannery dock back in port left me contentedly exhausted, and, in the end, certain in the knowledge that my youth was gone forever.

But those months provided the crucial funds necessary to allow me to continue the research for this book and to see its writing on to completion.

I am deeply indebted to the many commercial fishermen who openly shared their experiences with me, stories filled with adventure and bravery and panic, of impossible rescues and, all too often, death.

The names of their fishing vessels were: the *Amber Dawn*, the *Alaska Star*, the *Angela Marie*, *Bob's Boat*, the *Cloverleaf*, the *Gunmar*, the *Harder*, the *Margun*, the *Marysville*, the *Mia Dawn*, the *Neakanie*, the *Polar Star*, the *Rebecca*, the *Rondys*, the *Saint Patrick*, the *Ten Bears*, the *Tidings*, and the *M/V Wolstadt*.

In Juneau, Alaska, I would like to thank Debbie Schiedler Aranda, Greg Backman (USCG), David Beachem (USCG), Father Ron Dunfey, Shawn Elderidge, Mark Guillory (USCG), and Robert Newton for their assistance.

In Ketchikan, I wish to thank Marie Van Aarn, Pete Van Aarn, Shirley MacAllister, Tom and Dean Hordtvedt, Mik Sharp, and Danny Wall.

In Kodiak, I wish to thank Michael Barnes (USCG), Brian Blue (USCG), Lt. Larry Cheek (USCG), Nancy Freeman, Lt. Com. Richard Gaines (USCG), Thomas Goldston, Lt. Bill Gottschalk (USCG), Vern and Debbie Hall, Dr. James Halter, Joe and Mary Ellen Harlan, William De Hill Jr., Bruce Hinman, Gene Le Doux, Lt. Bob Lockman, Lynn Loftquist, Gary and Debbie Marlar, Shannon McCorkle, Captain Jimmy Ng (USCG) and his wife Joy, Dr. Martin Nimeroff (USCG), Hank Pennington, Bert Parker Jr., Pat and Dina Pikus, George Pikus, George Timpke, Chris Rosenthal, Paul Vines, and Wade Watkins.

In Petersburg, I would like to thank Norman and Mary Armin, Bill Beal, John Breezeman, Charlie Christiansen, Officer Rick Carter, Alann Erickson, Haftor Gjerde, Erik Kegal, Kurt Kivisto, Colyn and Carleen Lyons, Colt Lyons, Natacha Lyons, Myron Lyons, Bill and Carol Stedman, Mike Tolson, John and Bert Winther, Wayne Winther, and Matt the cook.

And I would also be remiss if I did not mention each and every one of the following people. Some are living; some are dead.

In alphabetical order, they are: Amron, Wayne "Worm" Baker, Ethel Bangert, Jim Batlien, Larry and Sharon Bennett, Ethan Bohannon, Frank Bohannon, Debra Brushafer, David and Jill Capri, Wink Cissel, Dale Dickenson, Jan and Denny Dimmitt, Mike Doland, Richard Fong, Sandy Fong, Alex and Laura Fong, Perry Fiscus, Rob Frazier, Bud Gardner, Gary Garland, David Graham, David and Nancy Green, Cornelius Green, Joyce and Wilburn Hall, Kim Handland, Jeff Hinshaw, Brian Hutchins, Mark and Martha Hutton, Gunner Ilhudso, Sr., Gunner Ilhudso, Jr., James Jobe, Tom Kaupenin, Robert Kidd, Gary Knudsen, John Lance, Dick Lawrence, Rick Laws, John and Kathy Ludahl, Mike Machleid, Craig McKay, Jerry Marston, Jerry Andrew Miller, Sherri Miller, Larry Murphy, Hilario Nevarez, Ronald Newton, Molly O'Neil, Bobby and Rachael Osborne, Wilson Pair, John Pappenheimer, Charles Parlett, Saimi M. Pesio, Don and Alma Quammen, Randy Ryker, Terry Sampson, Vanessa Sandin, Randy and Sue Scott, Art Simonton, David Sparks, Todd Stallings, Clifford "Doc" Steigell, Velda Sutton, Bob Swanson, Wallace Thomas, Lonnie and Pattie Waddle, Robert B. Walker, Tim and Linda White, Conrad Johnson and Maggie Wilhelm, Mel Wisener, Ed Wyman, and Fritz Youra.

Finally, I must thank Barbara Anderson and Cal Morgan,

the talented editors at St. Martin's Press with whom I was blessed to work; Jennifer Starrels, editorial assistant; Lance Rosen, my razor-sharp, Seattle-based attorney; and finally, Bart Eaton, who took the exceptional photo used on the cover of this book.

Introduction

F ar out on the brutal wilderness of the Bering Sea, 250 miles southwest of the fishing port of Dutch Harbor, another storm-battered king crab boat staggers toward home. But, burdened by the leaden tonnage of a two-foot-thick layer of sea ice encasing every inch of her superstructure, she rolls suddenly. Her disbelieving skipper sends out a Mayday. Eight terrified crewmen race to escape. Yet the 130-foot vessel flips over with astounding quickness, entombing the skipper and most of his crew inside. They struggle in the flooding darkness as the inverted ship descends into the crushing depths, carrying them to their watery graves on the ocean bottom over six thousand feet below.

Of the crewmen who are lucky enough to escape the murderous confines of the sinking vessel, just one, in all the panic, manages to grab his survival suit. When the hull sinks out from under the huddled, shivering souls balanced upon it, they are cast adrift.

Without the protection of a buoyant, insulated survival suit, the human body loses energy when immersed in water at twenty-four times the rate it does in air of the same temperature. The crewmen who fail to get hold of a suit are, in only minutes, chilled into unconsciousness by the heart-stopping 38°F. seawater and, one by one, drown. The lone survivor knows that, even wearing his suit, he will last only a few hours. And as he struggles to stop a leak in the chin area of his survival suit, thunderous storm waves three stories high collapse upon him.

Lifting his head to the howling black vacuum of night, the single survivor prays.

At the U.S. Coast Guard base back in Kodiak, some six hundred miles away, a radioman is quietly manning his post. Suddenly, his breath comes up short. He rocks forward in his chair, galvanized. He presses his headset to his ears and manages to intercept a brief but frantic Mayday. Then all contact vanishes. There is no way to pinpoint the foundering vessel's exact position, but he sounds the alarm and two C-130 U.S. Coast Guard Search & Rescue planes scramble. The four-engine, turboprop planes roar off down the glistening runway and disappear into the night.

Flying at a top speed of 250 miles per hour through the cold, wet darkness, it will take several hours for them to find the scene. Then they will have to circle and wait for daylight.

At first light, and using only the best technology of our day — the human eye — for spotting drifting survivors, the pilots will slow to 200 miles per hour at an altitude of five hundred feet and begin the search. Pressing on through the daylight hours, they will fly back and forth across a 5000-square-mile grid of trackless, wind-raked sea for any sign of life.

Brutal scenarios such as this occur each year to those who make their living in America's most dangerous profession: commercial fishing on Alaska's high seas. Now and then, a fortunate few live to tell of their harrowing experiences.

As one who has fished commercially in Alaska for more than a score of seasons, I find the stories told by the few men and women who have survived them, in a word, fascinating. I have spent several years trying to track down and interview a number of these survivors. The result is *Nights of Ice,* a collection of what I believe are some of the most exceptional stories of survival and disaster in Alaska ever assembled.

To the multitude of average working Americans caught up in the soul-crushing stress and icy indifference of their high-tech worlds, Alaska hovers on the horizon as a combination of mirage and panacea, a place where a spirited young man or woman can still go and roam free across a pristine wilderness; a place to which everyday people, fed up with their lives (and armed with nothing more than ambition and a strong back), might still escape to carve out a new beginning for themselves.

For me, Alaska was never a myth. In the twenty years since I first migrated to the port of Kodiak and landed a job as a young deckhand in the commercial fishing fleet, thousands more have also fled north and found work. And literally hundreds have died at sea. Forty-four perished in Alaska waters in 1988 alone.

But in the moody, iceberg-laden January waters of the Bering Sea, in the world's toughest fishery, where crab boats maneuver through white, drifting fields of polar ice in search of snow crab, a number of lucky deckhands recently earned $96,000 per man—in just thirty-one days of fishing.

Rags or riches, feast or famine, wealth, poverty, death: the fisherman who makes his living in Alaskan waters is a gambler at heart.

As young descendants of Horatio Alger wander north in search of opportunity, any one of the newcomers might strike it rich, retiring in just a few years, or he might die in a day. He might ride out decades of Bering Sea typhoons, only to be beaten to death as an old man, while fishing alone in a skiff, by the flailing tail of a two-hundred-pound halibut.

Often working around the clock, these men troll for salmon amidst the lovely, forested archipelago of islands that compose southeast Alaska. They battle powerful tide-rips that sweep across the entire volcanic thousand-mile length of the Aleutian Islands. In winter, in the savage waters of the Bering Sea, they commonly work on deck for forty hours without rest as twenty-foot seas toss their ships. And during the ice storms of winter on the edge of the polar ice cap, they work to exhaustion and beyond swinging baseball bats and sledgehammers to break off the deadly tonnage of spray-ice from the superstructure of their ships.

Regardless, this rugged and fiercely independent group of men and women ply more than a million square miles of untamed sea, working on the edge as they search for salmon, halibut, codfish, and crab.

There are 33,000 miles of shoreline in Alaska, more than the east and west coasts of the United States combined. And the richness and the vast diversity of her coastal regions are like no other place on earth.

NIGHTS of ICE

NIGHTS OF ICE

For Joe Harlan, captain of the fifty-three-foot crab boat *Tidings,* and his crew, the 1989 Kodiak Island tanner crab season had been an exceptionally tough one.

From the opening gun, they'd ignored the weather, fishing hard through the merciless cold of an arctic storm. They were working the waters down in the Sitkalidak Island area on the southeast side of Kodiak Island, some eighty nautical miles from the fishing port of Kodiak. But Harlan and his men had been pleased at their luck.

They'd been pulling gear in the biting cold of the short winter hours of light, grinding through a total of some one hundred crab pots. Harlan had agreed to pay his men a 10 percent per-man crew share that season. In just two weeks they'd boated more than forty thousand pounds of tanner crab. Crew shares had already topped ten thousand dollars—per man.

But they had been pounded night and day by wild northwest winds packing chill-factor temperatures of minus fifty degrees and williwaw gusts that made fishing that 1989 season one of the most perilous ever. One crab boat had gone down not far from Harlan and his men, near Chirikof Island. All four of the crewmen had died. So far, only one body had been recovered.

On another crab boat, one deckhand had lost three toes to frostbite when he ignored the water sloshing about in his boots while working on deck. As a fellow crewmate recalls it, "Before he knew it was happening, it had already happened."

On the very first day of the season, Joe Harlan had lost one of his own men to frostbite. He had rushed the man into the ancient Aleut village of Old Harbor on the south end of Kodiak Island and hired a bush plane to fly the man to the hospital in Kodiak for treatment. The man had broken no hard-and-fast rules of the sea; he had merely tried to sort crab wearing only cotton glove liners. And he had developed large blisters on the fingertips of both hands, which would keep him out of commission for the rest of the month-long season.

Less than two weeks of fishing later, the tanner crab catch fell off dramatically. And skipper Joe Harlan turned to his crew, hoping to cut his losses and call it a season. "Well, you know what you've made so far," he began. "And the way the crabbing has been going these last few days, you know what you can expect to make. The way I see it, we have two choices. We can stay out here and scratch away on a five-or-ten-crab-per-pot average until the Alaska Fish and Wildlife Department tells us to quit. Or we can quit burning up our fuel, store our gear away back in Kodiak, and get out of this cold son-of-a-bitchin' weather."

After weeks spent working in the single-worst extended

cold spell ever recorded in the Kodiak weather books, the crew of the *Tidings* did not hesitate. They'd been "successful enough" for one season.

Built in 1964 in the shipyards of Seattle, the fishing vessel *Tidings* had a wheelhouse that was mounted forward on the bow. She was considered one of the nicest boats around at the time because she had a toilet, something that was considered rather extravagant in those earlier, "hang it over the side" days.

Packing the recommended load of some thirteen crab pots on her back deck, the *Tidings* was closing fast on Chiniak Bay of the port of Kodiak, cruising through moderate seas, paralleling the coast of Kodiak Island about one and a half miles offshore, when, late on that frigid night, "all hell broke loose." Joe Harlan had been looking forward to slipping into the close and comfortable shelter of the Kodiak port, and with the exception of ice forming on the wheelhouse and railings of the *Tidings,* their journey north along the full length of the island had gone as planned.

But at Narrow Cape, they ran into some bad tide rips. Spray began exploding over the length of the ship, and they began making ice heavily.

Joe Harlan soon rousted his crew from their bunks.

"Guys, we've got to get this ice off of us," he said as he woke the men. With his crew gathered in the wheelhouse, Harlan pointed at the windows surrounding them. They were encased in ice. Only a clear space the size of a quarter in one window remained.

"Knock the windows clear, and then be sure and get the ice that's stuck to our railings," directed the skipper. "But don't let yourself get frostbitten. As soon as you get cold, come on in!" he insisted.

In an amazingly short time, a thick layer of ice had formed on the *Tidings.* Ice covered the boat—except, that is, for the crab pots themselves. Harlan and his crew had wrapped the commonly ice-drawing forms of steel and webbing in a layer of slick plastic, one that drained quickly before the spray from the ocean had time to harden.

The crew of the *Tidings* made good work of the ice-breaking task. But they longed to get back inside, out of the murderous cold. There they would flop down on the floor and warm themselves in front of the heater in the galley.

Back inside, skipper Joe Harlan was making a routine check of his engine room when he spotted seawater rising fast in the ship's bilge. Seconds later, the *Tidings* began to list to the port side.

Either the crab tank's circulation-pump pipe had ruptured inside the engine room or the steel bulkhead separating the engine room from the crab tank had split a seam. Nobody would ever know for sure. But it "put a lot of water in that engine room right now!" The moment Harlan spotted the water, he rushed to the back door and yelled to deck boss Bruce Hinman, "Grab your survival suits! And then start kicking the pots over the side!"

Joe Harlan was standing in the wheelhouse when he felt the *Tidings* roll. Instinctively, he was certain the ship would not be able to right herself. Yet he "couldn't believe it." He found he was unable to accept what was happening to the *Tidings,* a vessel he had come to trust and even admire. Then a ridiculous thought shot momentarily through his mind: He had an unspoken impulse to order his crew to "run back there, hop overboard, and push the thirty-ton vessel back upright."

With the *Tidings* sinking fast, Joe Harlan knew she would finish her roll and sink completely in about the time it would take him to utter a single sentence. Turning to his VHF and CB radios, he made a snap decision.

In the past, he'd listened to many Mayday calls to the U.S. Coast Guard in Kodiak. As glad as he was to have them standing by for his fellow fishermen, Harlan also knew that the Coast Guard usually wanted to know "who your mother's sister was, the color of your boat," your date of birth, your last checkup, proper spelling, and the like. So rather than shoot off a Mayday to the Coast Guard, Harlan decided to take a gamble.

For much of the night, he'd been listening on VHF channel 6 to the friendly chatter of the boats traveling ahead of him up the line. He grabbed the CB mike then and yelled, "Mayday! Mayday! Mayday! This is the *Tidings!* We're off Cape Chiniak and we're going down!"

As he spoke, the *Tidings* fell completely over on her starboard side. Below him, Harlan could see batteries breaking loose and flying across the engine room. He felt his heart freefall into his belly. Before he could unkey his mike, he "lost all power. Everything went dead."

Harlan heard a tremendous crash, and it seemed that all at once everything inside the boat—pots, pans, toasters, even rifles—came flying loose. Then the large hulk of the refrigerator came tumbling from its mounts. Harlan was thrown across the width of the wheelhouse. He struggled to regain his footing, but instead he tumbled backward down into the fo'c'sle.

Then, like a whale sounding, in one continuous motion the stern kicked high, and the *Tidings* sank bow-first, straight for the bottom. She slid toward the ocean floor in one steady motion, burying herself full length in the night sea. And there she paused, floating with only a few feet of her stern showing above the surface, with Joe Harlan still trapped inside.

As Bruce Hinman recalls it, shortly before the *Tidings* foundered and rolled over, his skipper had slowed the vessel to allow all six-foot-three and 290 pounds of Hinman's huge frame, as well as his fellow crewmates Chris Rosenthal and George Timpke, time to get dressed and make their way outside. A ten-inch-thick layer of sea ice had already formed on the *Tidings'* superstructure, and was growing fast. Clad only in their work clothes, they hurried outside to do battle.

Hinman, Rosenthal, and Timpke attacked the ice with baseball bats. They broke ice and tossed the chunks overboard as fast as they could move. And as they did, they squinted against the sharp, eye-watering gusts of arctic wind, and winced at the biting cold. They worked in drenching conditions in a chill-factor reading of some -55°F., the coldest ever recorded in the area. And they swung at the growing layers of ice now encasing the bow railings and bulwarks surrounding the wheelhouse, certain in the knowledge that their very lives hung in the balance.

Suddenly, the crab boat began to list sharply. The growing list soon tilted past forty-five degrees. Seawater rose over the port-side railing. Hinman was removing his survival suit from its bag when a tall wave broke over the twisting slope of the deck. And just as suddenly, he and the others found themselves dodging a deadly shuffle of 15,000 pounds of crab pots sliding forward toward them down the steep slope of the deck. As the bow of the *Tidings* nosed farther forward into the icy sea, the seven-ton stack of crab pots accelerated its slide, further distorting the already-untenable balance of the sinking ship.

Accelerating as it came, the tall and deadly stack of sliding crab pots closed on the terrified crewmen like a moving mountain. The square-fronted stack of steel and webbing plowed into the back door of the ship's wheelhouse like a runaway freight car. It slammed against the rear of the wheelhouse with the effect of a door closing on a bank vault, leaving their skipper trapped inside.

With the shifting weight accelerating the angle of the plunging bow, the *Tidings* rolled with an astonishing velocity, pitching the four crewmen scrambling across her back deck bodily through the air and overboard.

The suddeness of the port-side motion caught everyone off guard. It was as if an all-powerful force had suddenly gripped the *Tidings* and flipped her—as if her fifty-three feet and forty tons were no more substantial than a bathtub toy in a child's hands.

Bruce Hinman felt the sudden shift, and he found himself hurtling through space; several of the crab pots followed. His right arm became entangled in the webbing of one of the pots—and just as suddenly, the six-hundred-pound crab pot began to "sink like a rock" toward the bottom, dragging Hinman, kicking and struggling, along with it.

It all happened so quickly. Hinman had been knocked senseless by the sudden shock of the Kodiak waters, ensnared by one of his own crab pots, and was now being dragged along on an unforseen journey into deepest darkness toward an ocean floor more than a thousand feet below.

He knew instinctively that if he allowed panic to rule him, he would be lost. And he fought to choke back the rising tide of unreasoning fear within himself.

As he descended through the darkness, Hinman gained a measure of composure. He would fight against the building fear by taking action. He was perhaps seventy feet beneath the ocean surface when he managed to jerk his ensnared right arm free. Then he placed both of his stocking feet against the webbing of the crab pot and pushed away violently. The fast-sinking crab pot disappeared quickly, tumbling off into the black body of sea below him.

When Hinman looked up, he was awestruck by what he saw. For a blinding orb of radiant light hovered above him. There was something beautiful, even angelic about the vision before him. Brilliant in splendor, it bathed him in spirit-lifting columns of golden light that seemed to beckon him home.

He ascended feverishly then, stroking overhead toward the comforting swath of inexplicable light like a man with a building hope, a hope tempered by the fear that at any moment another toppling crab pot might very well descend upon him and carry him back down again.

And at that moment, Bruce Hinman's past life flashed before his very eyes. Launched instantaneously through time, he watched the events of his life play out before him with "the speed of thought." The prevailing feeling was of being cast adrift on a wondrous journey, unhindered by earthly impediments of time, matter, or communication.

Hinman felt "lost in time without an anchor." And the look and feel of special moments long past came back to him now with complete clarity. They flashed and froze there in his consciousness, in a kind of nostalgic collage of all that had once mattered in his life.

He saw his two little boys; his former wife, Carol; his two adopted foster daughters; and both of his parents, as well. Then Hinman was back under fire in Vietnam, just as it had all happened with soldier buddies dropping all around him. A millisecond later, he was a boy again, scrambling along the banks of Lake Shasta in northern California. And he was swept back into the very moment when he had come so close, as a child, to drowning. It all scrolled past him now, and each memory carried with it the exact same heart-tugging emotion he had felt at that time.

Bruce Hinman exploded through the surface, leaving the visions behind. He rose bodily into the bitter night, then wrenched hard and began inhaling deep lungfuls of the precious air.

When he regained himself, he spotted the stern of the *Tidings* drifting nearby. She was hanging straight down in the water.

Adrift now, without a survival suit, lost in a whiteout of silvery gray ice fog, Hinman knew the odds of outliving his predicament were slim. He dog-paddled and fought to catch his breath.

When he could, he yelled for his crewmates. "Hey, Chris! Where is everybody?"

A voice sounded out of the darkness perhaps fifty feet away. "Hinman!" He recognized the voice of his good friend and crewmate Chris Rosenthal.

"Harlan's in the boat! He's down inside the boat!" Chris shouted.

George Timpke, their third crewmate, soon acknowledged him, as well. He was clinging to a piece of flotsam off in the darkness approximately one hundred feet away.

Short of diving equipment, Hinman knew he had no way of reaching his trapped skipper. Swearing aloud, he soberly acknowledged to himself that his good friend Joe Harlan was a goner. And a single thought shot across his mind. What am I going to say to his wife, Mary Ellen?

Now Hinman felt the almost caustic effects of the bitter wind and numbing ocean against the flesh of his face. He also knew there would be no way to climb back aboard the sinking vessel, no way for him or his other shipmates to climb clear

11

of the life-sucking cold of the Gulf of Alaska water, and he felt at once helpless and angry.

"Now what?" he shouted into the arctic night.

Harlan had been roughed up considerably when the *Tidings* had rolled over. In fact, he had come close to breaking his right arm. He scrambled to gather himself and climb out. Ordinarily, the ladder leading from the engine room to the sleeping quarters stood upright and led down into the engine room. But now the vertical leg of the ladder posed a serious obstacle. With the *Tidings* tilted straight down as she was, the ladder now lay unevenly across the inverted space before him, sloping as it stretched between cabins. Scaling it would be a little like trying to climb the underside of a stairway. With his battered right arm, it would be a difficult gymnastic feat.

Now the startled skipper found himself "all the way forward" in the darkest inner reaches of the ship's bow, some fifty feet below the surface of the sea. As the boat continued to leap and roll, he could hear the ongoing crash and clutter of stored parts falling and scattering overhead.

Suddenly, in the gray-black light, a roaring blast of seawater broke through the door to the engine room. The tumultuous white water broke heavily over Harlan, lifting him bodily and washing him out of the fo'c'sle. He gasped for air as the icy flood cascaded over him. As he was carried along through the inverted space of the ship's galley, Harlan reached out and snagged the handle to the wheelhouse door. He tugged frantically, but with the water pressure sealing it shut, he found it immovable.

As the small galley continued to flood, Harlan found himself struggling to remain afloat in the narrowing confines. The galley sink and faucet were now suspended on end below him while beside him, in the claustrophobic space, floated the gyrating hulk of their refrigerator.

Harlan gulped air and dived. He knew he had to think of a way out. He swam down through the watery cubicle of the galley to the sink, grabbed the faucet with both hands, and kicked viciously at the starboard side window behind it. But the leaden cold of the water seemed to drain the power from his blows. This is hopeless! he thought, as he swam numbly back toward the pocket of air above.

The moment his head emerged, he was greeted by a terrifying roar. The flood of gushing seawater into the room seemed to be accelerating. The sound of it echoing in the small sliver of space was deafening. The water sloshed back and forth between walls that rolled and dipped in a dizzying motion around him. Again, the refrigerator drifted into him. And he fought against a building sense of horror.

Treading water, Harlan tilted his head back in the narrow space next to the ceiling and tried to inhale the precious air. But the shocking cold of the seawater continued to make breathing difficult, and his breath came in shallow huffs.

Well, this is it, he thought. This must have been how Jim Miller died on the *George W.*

Harlan considered praying, but it occurred to him that doing so would be an admission that he was going to die. He also made up his mind "not to snivel." He would not pray, and he would not blubber. He would face the outcome, whatever that might be.

Joe Harlan weighed the chances of escaping through the galley door and out onto the back deck. But with the *Tidings* standing directly on end as she was, he knew the entire

15,000-pound stack of crab pots would be pressing down against the door at that very moment. And, with the refrigerator floating in front of it, he conceded the escape route had been lost.

Yet even at the time, in the midst of all the terror and commotion, Joe Harlan realized that there was something strange about his ongoing ordeal. For he could see virtually everything. With his adrenaline flowing, his senses had somehow become heightened — and now a whole new world seemed to open up before him. When he dived again, he saw, through the blue-green tint of the water, the forms of the faucet and the window behind it, while the blocky brown figure of the refrigerator bobbed above him, suspended in the water overhead.

There was something strange about it all. Harlan knew there were no lights burning on the boat. Everything had gone dead. There was "zero power." Perhaps it was moon-bright up top. But then he recalled that there had been no sign of the moon on such an inclement night. Yet, submerged as he was, Harlan could see clearly through the seawater inside the boat, as well as out into the light green sea space on the other side of the window.

Adrift in the ocean current, the hull of the *Tidings* bounced now in the lumpy winter seas like a floating berg of ice, with barely 10 percent of her whole self still showing above the surface.

Entombed inside the sinking hull of the *Tidings*, Joe Harlan was also feeling the weight of his impossible predicament close in on him, and his emotions built toward a breaking point.

He thought of his lovely young wife, Mary Ellen. She had soft brown hair, and beautiful blue eyes. They had met four years before in Kodiak; Harlan had hired her to do the cooking for the boat during a herring season. They had married soon after that, and now they were the proud parents of a beautiful one-year-old daughter, Chelsea.

In the hardworking and yet contented years since then, Joe Harlan and his wife had built a fine house together outside of Kodiak in the Basket Flats area along Sergeant Creek. Harlan could see the ocean from the balcony of his home. And he had one of the best silver-salmon fishing holes on Kodiak Island right in his own backyard. During the salmon-spawning season in the lush and beautiful summer months on the island, he often had to put up with Kodiak bears who wandered onto his backyard property and competed for those same spawning salmon.

Joe Harlan loved his family. Besides, he had mortgage payments to make, and a lot of living to do. The whole damned thing just wasn't fair. He couldn't bear to accept it! He couldn't "just give up!"

The injustice of the moment sent Harlan into an emotional spiral that carried him over the edge. And he erupted into a blind rage. Wild with anger and determination, he sucked in another brief pull of air and dived again for the sink. He would make another attempt to break out the window. But this time, he would try another method. He swam to the sink, then grabbed the faucet tightly again with both hands and began repeatedly ramming his head into the glass.

Suddenly, the window exploded from its mounts. Harlan watched as it tumbled out into the pale green void and fell into the watery oblivion below.

All of a sudden, Harlan felt outside of himself. Imagining

15

himself to be a sea otter, he swam nimbly ahead through the small opening as if it were the most natural thing, arched his back, and headed directly for the surface.

In the strange and unexplained illumination that still remained, Harlan was able to see the hull of the *Tidings* as he swam upward. Man, I can't be *that* far from the surface, he thought as he stroked "up and up and up." The moment he broke through the surface, he felt himself return to his old self. It was like breaking into a "completely different world again!"

Harlan gasped wildly for air.

B ruce Hinman was drifting next to the bobbing stern of the *Tidings* when a man's head exploded through the surface, popping up right alongside him.

Choking, thrashing against the water, the man coughed heavily and spun in his direction.

"Hinman! You ugly son of a bitch!" he yelled.

It was none other than his skipper, Joe Harlan.

"Joe! Damn, I thought you were dead!" shot back Hinman, elated to see him.

Hinman's levity at seeing Harlan was quickly tempered, however, by the hopeless realization that there was no way to survive the present predicament. The canister containing the life raft had apparently failed to release when the *Tidings* rolled over, or perhaps it had released, only to get tangled up in the rigging or the crab pots. It didn't matter. Without that life raft, they knew they were all "as good as dead."

"What are we going to do?" shouted Hinman to Joe Harlan. "Did we get off a Mayday call?"

Harlan had tried, but he couldn't be positive that anyone had heard it.

There was nothing the men could do now but tread water and wait. With the wind blowing offshore, there would be no way to try to swim to shore. The deadly effects of hypothermia commonly paralyzed and drowned most men adrift in such seas in a few short minutes, at least those wearing only work clothes. Some began to sink the moment they hit the water. Even if Harlan, and Hinman, and the others could remain afloat, the wind and waves would eventually carry their bodies out to sea, where they would be lost forever.

When one crewman realized how grim things looked, he announced that he might just as well swim back down into the wheelhouse of the *Tidings,* get his pistol, and shoot himself.

Suddenly, an object "as large a dinosaur" exploded out of the water between Hinman and his skipper. It was the fiberglass canister that housed the *Tidings*'s life raft. The canister was about the size and shape of a fifty-five-gallon oil drum.

"Grab that SOB!" yelled Hinman.

The four crewmen converged on it. "Pull the cord!" yelled Joe Harlan.

While his crewmates treaded water nearby, Hinman began to peel off the line as fast as his numbed arms could move. He pulled and pulled, and after what seemed like several hundred feet of line later, he aired what everyone was thinking.

"God Almighty! This must be some kind of joke! We've got nothing but a coil of line here!"

"Pull! Pull faster!" yelled one terrified crewmen.

"Damn, man! I'm pulling as fast as I can," shot back Hinman.

Moments later, Hinman was forced to stop. The frightening

cold was beginning to press in on him, and he'd run out of breath.

Joe Harlan soon rejoined Hinman in the effort. He pulled what seemed to be literally hundreds of feet of line from what they had supposed to be the life raft canister. It was beginning to look like a fisherman's prank—an unbelievably cruel prank—had been played on them. What if the white fiberglass canister floating in front of them, the canister designed to house the ship's life raft, was filled with nothing but a large, unending, useless coil of line?

Minutes dragged by as Hinman continued to extract the line. As hundreds of feet of useless and entangling line played out in the water all around the floundering crew, their worst fears began to take on the feel of reality. When they came to the end of the line, a winded Bruce Hinman wrapped the line around his numb, pain-racked hands and gave a final tug.

Nothing happened.

The crew treading water around him let out groans filled with disbelief and a mounting panic.

Harlan's mind raced. I just can't believe this is happening! he thought. After all we've been through, to come up short like this.

Joe Harlan moved in to help. He leaned back in the water, placed both feet on either side of the end of the canister, and wrapped the rope line around both of his clumsy, cold-ravaged hands. He reached down all the way then and pulled with everything he had.

The stubborn knot on the other end of the line gave way suddenly. Then came the pop and hiss of the CO_2 cartridges discharging within. In the next instant, the bright orange canister exploded open and the raft began to inflate. But it inflated upside down.

As longtime fishermen, 290-pound Bruce Hinman and 200-pound Joe Harlan continued to work together. They quickly assessed the situation and, without comment, approached the task at hand as if driven by the logic of a single working mind.

Inflated and upright, these rafts are fluorescent orange in color and round in form, with a diameter of eight feet. Floating, they look like giant inner tubes, or perhaps like those inflatable, backyard pools that small children use—with a dome tent mounted on top.

Swimming to one side of the raft, they crawled atop it. Then, planting their feet (and combined weight of five hundred pounds) on the downwind side of the overturned raft, they reached across, grabbed its upwind edge, and lifted it in unison. When the twenty-five-knot winds caught the exposed upwind edge of the raft, it flipped it upright, scooping more than a foot of icy seawater along with it as it did.

"All right!" yelled one shivering crewman as he breast-stroked nearby.

Drifting in the murderous cold of the ocean currrents, the entire crew was thoroughly chilled, their movements sluggish with the steadily advancing effects of hypothermia. Hinman and Harlan decided to drift alongside the raft in the painfully cold seawater and help their crewmates crawl aboard through the narrow doorway of the raft.

Being by far the huskiest of any man in the *Tidings* crew (or in the entire Kodiak crab boat fleet, for that matter), Hinman insisted on going last. It was a wise decision, for when all had been helped aboard, so numbed was he, and so completely had his strength been sucked from his body, that it took not only all of his own failing strength but also the body-wrenching efforts of the entire crew to haul him aboard.

Never in more than a century of brutal Alaskan winters

had a storm front this cold struck the Kodiak Island area. A -40° F. reading in Alaska's dry interior country near Anchorage or Fairbanks was considered cold, even dangerous, although not unusual. But it was unheard of in the moist marine waters of the Gulf of Alaska.

The storm winds howled incessantly. The unrelenting gusts turned the raft's doorway into a virtual wind tunnel. Caught without a single survival suit among the four of them, and constantly awash with more than a foot of icy Gulf of Alaska seawater crashing about inside their raft (and with more seawater washing inside all the time), the crew of the *Tidings* knew their lives were still in serious jeopardy.

In truth, the record cold front threatened to freeze them where they sat. Packed tightly inside the cramped and drenching confines of the dome-covered raft, the cold-ravaged crew of the fishing vessel *Tidings* huddled together, shivering violently as panting columns of steamy breath jetted from their mouths.

Bruce Hinman rubbed his hands together furiously. He crossed his forearms, folded his hands under his armpits, and turned numbly to his skipper.

As if the record cold and unconfirmed Mayday hadn't been enough to worry the crew of the *Tidings,* they now discovered another unsettling fact: Their painter, leading out from the life raft, was still tethered to the sinking hull of the *Tidings.* In theory, the raft was attached this way to keep the crew members in the close vicinity of the boat as long as possible. But the status of the inverted *Tidings* was tenuous at best, and they knew she could be heading for the bottom at any moment. If she did, the life raft and all its occupants would likely be pulled down along with her.

The bridle cord attaching the raft to the painter was

designed so that it was tethered directly in front of the raft's entrance hole. In any windy conditions, this meant that as long as the raft remained tied to her mother ship, the gaping hole of the doorway would always end up facing directly into the prevailing wind. That wind now drove close-cropped ocean waves against the side of the stationary side of the raft. And icy walls of sea spray began exploding in through the front door and over those inside. Short of cutting the cord and casting themselves off into the mercy of the night, there was nothing to be done.

In only minutes, the blunt force of the record cold, the knifing edge of the arctic wind, and the drenching blasts of icy sea spray had rendered the men almost unconscious. They prayed then, and waited. And as the murderous cold bore down on them, a heavy silence fell on the crew.

Like the rest of the men, skipper Joe Harlan could no longer feel his fingers. But when he sensed the growing sense of hopelessness in the raft, he turned to his men.

"Look, guys, we're going to make it. Try not to worry about it. We're in the raft. That's the important thing." He paused. "We've just got to keep fighting it," he added. And he set about to keep the men busy. "Now is the time to get things done. And the first thing we need to do is to get that door flap tied shut!"

Those nearest the door opening soon discovered that the flap ties were frozen fast to the walls of the domelike ceiling of the raft. The going was slow and painful. No one in the entire crew seemed able to carry out the simple task; their numbed fingers had lost all dexterity. Yet if the crew was going to survive, it was imperative that someone tie the thin strips of nylon fabric to the bonnet of the raft itself and shut out the deadly chill of wind and sea. Harlan encouraged them to keep trying.

With the rest of the crew now on task, Hinman and Harlan worked frantically to open the survival kit. Perhaps there was a knife inside that would allow them to cut themselves free of their mother ship. They soon came upon the small package containing the essential lifesaving equipment such as flares, water, and food. But whoever had packed the raft had wrapped the package in layer upon layer of silver duct tape. Joe Harlan discovered that his fingers were no longer taking messages from his brain. His fingers had given out, and so Harlan began attacking the wide silver-gray tape, ripping at it with his teeth.

One crewman started praying again. "Dear God, help us! Dear God, help us!" Over and over he repeated it.

At first, Harlan appreciated the prayer; then it began to wear on him. With the icy weight of the Alaskan cold front bearing down heavily upon them, the freezing crewmen were "starting to fade." Finally, Harlan spoke to the crewman. "You know, you need to shut up now," he said steadily. "This is not good for our morale." The young crewman fell silent.

"Guys, we're going to make it. We're going to make it. So let's just keep thinking that way," Harlan added.

Joe Harlan was proud to see how his crew worked together. In a situation where fatalism might not have been out of line, they were doing all they could to save themselves. There isn't a coward in the bunch, he thought.

When they finally got the survival package open, Hinman and Harlan found no knife; they did manage to locate a small flashlight, yet its batteries had lost most of their charge. And they were forced to squint hard in the dim and intermittent flashes of light to read the flare instructions and figure out how to work them.

Harlan held out one of the flares and turned to Hinman. "Bruce, how do you work one of these damned things?"

Hinman looked at the oddly constructed flare. It had foreign instructions printed on it. He handed it back to Harlan.

"The goddamned instructions are in French!" shouted Hinman. And he cursed a streak.

Unable to read the instructions, and fearing they might accidently launch the flare into the face of someone inside the raft, Joe Harlan decided that for the time being he would not attempt to launch one at all. Joe Harlan now felt himself slowing down dramatically. And each time the cord line leading from the raft to the *Tidings* pulled tight, the life raft would once again contort wildly beneath them, and the severely hypothermic men inside would be drenched in yet another icy blast of seawater. It soon grew so cold inside the raft that the men agreed that they'd felt warmer while immersed in the sea itself.

Finally, after herculean efforts, they managed to get the flap tied shut. But even then, it only partially blocked the painful, drenching blasts of exploding sea.

Adrift in the cold and utter darkness, they remained tethered to the bouncing, drifting hull of the *Tidings,* blown back and forth across the rugged face of the sea by a knifing twenty-five-knot wind and battered by an unforgiving sea. With nearly a thousand pounds of men sprawled on the floor inside, Harlan felt certain that the constant jerking of the waves would soon tear the raft in two.

Harlan knew what he had to do. He searched for his own knife. He felt a lump inside one of his pockets, reached in one pocket, and pulled out a stray shotgun shell. Finally, he managed to locate his knife. But when he ordered his hands to open, they refused. Then, holding the knife in the palm of his stiff hands, he bent forward, took the edge of the steel blade in his teeth, and pried it open. Leaning outside through the doorway of the raft, he slowly and deliberately sawed on the line.

The line parted, and suddenly they were adrift. The violent, neck-snapping action of the raft vanished abruptly. Now as they rode up and over the rolling seas, they could hear the roar of the wind and the breaking of unseen waves off in the darkness all around.

Then a thought came to Harlan: Hypothermia isn't a bad way to die. After the initial cold goes away and you go numb, you just start slowing down. Pretty soon, you get lethargic, and you just feel like going to sleep.

He fought against the seductive nature of such thoughts by admonishing himself. "Don't you give up! Don't you go to sleep, now!"

Bruce Hinman furiously rubbed his hands together. He crossed his forearms and folded his hands under his armpits. Then he pulled the door flap a few inches to one side and scanned the late-night seascape all around.

The sinking of the fishing vessel *Tidings* brought with it an especially insistent message. This was the second ship to sink out from under Hinman in the last month. Both had sunk off that very same point of Kodiak Island coastline—Cape Chiniak.

The U.S. Coast Guard squad, flying out of the Kodiak Island base, had been kept hopping all season long. The rescue of Hinman from the sinking *Cape Clear* several weeks before had been performed in huge seas in yet another blinding snow-storm.

The Coast Guard helicopter pilot had descended bravely out of the night and hovered down over the sinking vessel. But then the helicopter's rotor blades had struck the ship's mast, very nearly killing Hinman as well as the eight men on board the chopper.

Adrift then in the tall seas, Hinman had fought hard to keep from drowning as the torn and flooded suit he wore threatened to sink him. He was completely played out by the time they finally managed to hoist him aboard the Coast Guard chopper.

And now he and his crewmates were waiting to be rescued from yet another crab boat. Hinman was staring out through a silver moonlit haze of ice fog swirling across the lonely black face of the sea, when he spotted a set of approaching mast lights.

"Hey!" he yelled aloud. "Here comes a boat!"

The fifty-eight-foot fishing vessel *Polar Star* had been under way several miles off Cape Chiniak when the *Tidings* first called for help. The skipper and owner of the *Polar Star,* Pat Pikus, was wrestling with poor visibility himself at the time. He had been standing alone at the helm, moving ahead through a steamy, boiling cloud of ice fog, when the call for help suddenly leapt from his CB radio: "Mayday! Mayday! This is the *Tidings!* We're off Cape Chiniak and we're going down" Then, just as suddenly, the frantic voice fell silent.

Pikus quickly awakened his crew. "Everyone get up right away!" he yelled. "We've got a problem!"

He paused while his crew scrambled to life. Knifing, thirty-knot winds, with a bladelike edge of -26°F., were driving across the face of the sea. More important, Pikus knew there were no charts that could adequately describe the chill factor—nor the utter aloneness a drenched and drifting crew would know on such a night. When crewmen Shannon McCorkle, George Pikus, Gene LeDoux, and William De Hill, Jr., had gathered in the wheelhouse, he turned to them. "Boys," he said, "we've got a boat in real trouble nearby us here. And cold as it is outside, I'm still going to need one of you men to go climb up on the flying bridge and keep a watch out from there."

The wind was blowing offshore at the time, and Pikus began his search by making passes back and forth across the brackish water between the shoreline of Kodiak Island and an imaginary point several miles offshore. He had no sooner begun his effort when another skipper's voice jumped from the radio.

The skipper claimed that the last time he'd seen the *Tidings*, she'd been cruising several miles offshore. Still another skipper added that he believed he'd seen a tiny blip on his radar screen in the very area where the *Polar Star* was now cruising. But his radar had only fastened upon it once; then it had disappeared, and had never shown again.

After completing several grid-line sweeps, Pikus was about to head back into shore for yet another pass when, squinting through the boiling fog, he thought he saw something dead ahead. It turned out to be the silver flash of a small piece of reflector tape and it was stuck to the side of the dome of a life raft.

Slowing his approach, Pikus and his crew soon spotted the stunning figure of the *Tiding*'s stern bouncing slowly and rhythmically through the choppy black seas. The *Tidings* had somehow managed to remain afloat, standing on end, with almost her full length buried beneath the sea. Only the last few feet of her stern and rudder now showed above the surface.

As he watched, the exposed stern of the wave-slickened hull performed an eerie ballet. What remained to be seen of her rose and fell through a jet-black world of swirling fog and howling wind, a void as cold and oppressive as a journey into the unlit bowels of a walk-in freezer.

Pikus was afraid that, in the strong winds, his vessel would drift right over the top of the life raft. So he swung in downwind of it, then maneuvered in close.

"Hello! Hello! Is anyone there?" Pat Pikus yelled out his side wheelhouse door.

A muffled cry came back. Then the door flap on the side of the raft's dome flipped out and someone yelled, "Yah, we're here!"

The raft was caught in the bleak glare of his sodium lights. When he pulled along side, Pikus "looked right down into the raft." He had never seen a more pathetic sight. "No one wore survival suits," he recalls. "A couple of them were without shoes. There was a lot of water slopping around in the raft." The entire crew looked as weak and hypothermic as humans can get and still remain alive. "They wouldn't have made it another ten or fifteen minutes," he recalls.

By the time the *Polar Star* came abreast of their raft, Bruce Hinman was barely conscious and completely unable to stand. The crew of the *Polar Star* climbed overboard and literally dragged him from the raft, up and over the side, and aboard their ship. Hinman remembers landing on his back and the icy crackle of his sopping-wet clothing freezing instantly to the deck.

Joe Harlan reached up and tried as best he could to grab ahold of the railing. When they saw that he, too, was unable to be of much help to himself, the rugged young crewmen aboard the *Polar Star* reached down and, in one motion, hoisted him up and over the side. They tossed him onto the deck and out of the way in order to make room for the rest of the survivors.

Lying on the deck where he landed, Harlan spotted the door leading into the heated space of the *Polar Star*'s galley. Unable to walk and unwilling to wait, he rolled over onto his stomach and began crawling toward the door. Pausing en route, Harlan gathered himself, and, raising up on one elbow, took one last glimpse at what remained of the *Tidings*. Waves were exploding off the few final feet of her stern.

"Good-bye, girl," he said aloud. Then he collapsed back down onto the deck and began crawling again toward the warmth of the ship's heated interior.

So intent were they on rescuing the other survivors that no one among the ship's crew noticed Harlan go. He managed to crawl in through the galley, down the hallway, and into one of the staterooms, where he pulled himself "up into some-body's bunk" and lay there "shivering violently."

"Don't let them go to sleep! Keep them awake," ordered the Coast Guard repeatedly over the radio set.

When the crew of the *Polar Star* found him, Harlan peered up at them with dark sunken eyes from the soft, warm bunk in which he lay.

"Look," he said, "I want you to know that I'm married. And I've got a kid. And I don't want you to think I'm a homosexual or anything. But I need someone to take off all his clothes and climb in bed with me here. Because if you don't, I think I'm going to die."

It was *Polar Star* deckhand Shannon McCorkle, son of well-known Kodiak harbormaster Corkie McCorkle, who did the honors.

"He was the one who brought me back to life," recalls Harlan gratefully. "The real heroes of this thing were the crewmen of the fishing vessel *Polar Star*," claims Joe Harlan. "There's no doubt in my mind. If we'd been out there even another fifteen minutes, we would have died. We were that close to buying it."

By the time they managed to lift aboard the nearly frozen crew of the *Tidings*, nearly a foot of ice had accumulated on the decks and superstructure of the *Polar Star*. The instant the last man arrived on board, they left the raft to drift, and immediately struck out for Kodiak.

When the *Polar Star* arrived back in town, there was an ambulance waiting for them, but Harlan wanted nothing to do with the hospital. "Look," he told the EMTs, "I want you guys to take my crew to the hospital. Have them checked out and make sure they're okay. But I'm going home to see my wife and my daughter."

Throughout the entire ordeal, Harlan knew that it was the love of his wife and daughter and home that had kept him going.

Now barefoot, his wet hair still matted against his head, Joe Harlan was clad in nothing more than a wool blanket when a friend drove him home. His wife came out to greet him. It was a tearful reunion.

That winter, during the bitter cold of the crab season, Joe Harlan had grown a beard. Now even his one-year-old daughter did not recognize him. When he approached and picked her up, she asked him, "Are you Santa Claus?"

Once inside, Harlan took a long, hot bath, devoured hot platefuls of food, and spent time relaxing with his wife and daughter. At 8:00 A.M. the very next morning, Joe Harlan called a ship broker in Seattle. It was time to start shopping for a new crab boat to buy.

CHOPPER RESCUE: MEN IN PERIL

The skies over Kodiak Island broke cold and clear on that wintery February morning in 1985. Arctic storms had already dumped more than three feet of snow on the village of Kodiak that month alone, during one of the worst winters in Alaskan history.

But the bitter arctic weather brought with it not only a frigid, biting cold; for U.S. Coast Guard H-3 helicopter pilot Capt. Jimmy Ng (pronounced *ing*) and his crew, it brought a growing sense of urgency, as well. A hunter had become lost and was reported missing overnight on nearby Afognak Island. The Coast Guard had sent Ng, his chopper, and crew to look for the overdue hunter. At first light, Ng immediately initiated a grid-pattern search and soon spotted the man. He was cold and shaken, but grateful.

A short time later, during that same remarkable day, Ng

and his crew were involved in another emergency call, and were in the middle of hoisting a heart attack victim off a boat in an inlet on the west side of Kodiak Island, when Ng received a radio call from the Operations Center at his home base back in Kodiak.

Coast Guard operators, he quickly learned, had picked up a frantic Mayday call from the captain of a crab boat caught in a fierce ice storm near Cold Bay, out in the vast and primitive wilderness of the Alaska Peninsula.

The name of the vessel was the *Mia Dawn.* A fifty-eight-foot limit seiner, she had reportedly run aground on a semi-submerged pinnacle of ocean rock and was taking on water. Caught in subzero temperatures in total whiteout conditions, the ship was also icing badly. Huge seas were breaking over her entire length. Drenched and freezing, her three-man crew was trying to pump up the vessel's leaky life raft in what were reported to be howling ninety-mile-an-hour winds. The captain was pleading for someone to assist them.

Ng was ordered to return to his base at the earliest possible time. His luck was holding: The airlift of the heart attack victim came off "without a hitch," and he hurried back to Kodiak at top speed. He landed on the ramp and transferred his patient to a waiting ambulance. Then he "hot-refueled," an emergency technique in which a helicopter is left running while it is simultaneously off-loaded, resupplied, and re-fueled.

Moments later, the H-3 helicopter rose into the cold, rarified air and roared away. "We took off from taxiway Alpha with a turnaround time of less than ten minutes," Jimmy Ng recalls. Flying at the highest speed they could sustain, Ng and his copilot, Lt. Com. Larry Cheek, set a direct course for the terrified crew of the *Mia Dawn.* Also on board was Dr. Martin Nimeroff and flight engineer Michael Barnes.

The Huey helicopter — primitive by today's standards — had a respectable cruising range of six hundred miles under favorable weather conditions. The *Mia Dawn,* however, was approximately 440 miles away.

In perfect weather, the aircraft was capable of cruising at 140 knots. But they soon found themselves battling strong headwinds, and they knew that their actual ground speed, as well as fuel supplies, would suffer substantially because of this. It would take several hours to reach the general area of the sinking vessel. No one could estimate how long it would take to locate the vessel or how much time the rescue itself might take.

Shortly, a four-engine, propeller-driven C-130 SAR (search-and-rescue) airplane was also launched from the U.S. Coast Guard base in Kodiak. Its mission was to sprint ahead at 270 miles per hour, make visual contact with the sinking vessel if possible, and remain in the area, circling and monitoring the situation, until Jimmy Ng and his crew arrived on the scene.

The C-130 aircraft soon passed Ng's helicopter overhead. As the airplane neared the Semidi Islands, far from the visual comfort of Kodiak Island, the wind increased sharply. Then, as the C-130 closed on the rescue site coordinates of the *Mia Dawn* over on the Alaska Peninsula, the pilot began to encounter what he described as "truly exceptional" winds. The nearer they drew to the rescue site, the stronger the arctic winds blew.

*   *   *

Throughout the early-morning phase of the H-3's flight, the weather remained clear and cold—good traveling conditions except for the strong and ever-building headwinds driving into them.

Along the west side of Kodiak Island, a heavy rain began to fall, reducing visibility to between two and five miles. Then the helicopter's all-important radar screen blinked off. Ng's helicopter would be without radar assistance for the duration of the mission.

The weather was rainy and "ducky." When they reached the area of Sitkinak Strait on the south end of Kodiak Island, they started picking up a lot of wind and were forced to turn and maintain a constant in-flight "crab" of about fifteen to twenty degrees en route to the scene. (A "crab" occurs when a chopper flies with its nose facing into the wind and its tail cocked to one side in an effort to compensate for too much wind; the motion looks something like a scampering crab as it angles across the ocean floor.)

About halfway across "the pond" (that wide body of ocean water between Kodiak Island and the Alaska Peninsula), the temperature outside plunged to about zero degrees and the wind picked up to about forty-five knots. Then it began to sleet and snow. Forced lower by the weather, Ng flew all out at about three hundred feet above the white, collapsing crests of the wind-raked seas. Periodically, the weather closed in tighter, and visibility was reduced to about only half a mile. When that happened Ng chose to fly even closer to the water.

They were "crossing the pond" when a second essential piece of navigational equipment—their loran computer— failed them. The rains and wind they'd encountered en route had been so heavy that water leaking in through the helicopter's radar system had drained down into the loran, knocking both instruments off-line. From now on, Ng and his copilot would be forced to navigate by the seat of their pants. Unless their radar or loran came back on-line, they would have to conduct the search for the *Mia Dawn* either through visual contact or by the use of their DF (directional finder), a primitive device that was nothing more than a needle mounted in an instrument-panel gauge that could home in on radio waves. But if the ever-worsening weather was any omen, the rescue mission under way could very well prove to be impossible without a radar or loran computer working on board.

Now, as they neared the Semidi Islands, the turbulence increased sharply. But due to the desperate plight of the crew of the *Mia Dawn,* Ng elected not to go around it. Onward they flew, taking as direct a course to the rescue site as possible.

Several hours into the flight, Ng spotted something in the distance. Looking far out through the haze and across the water, he could see tall ocean waves exploding high and white against what looked to be a pinnacle of rock sticking bolt upright out of the surface of the sea.

Stationed at the Kodiak base for nearly a decade, Ng was one of the most experienced U.S. Coast Guard pilots in Alaska. He had completed countless rescue missions over that same remote sea route. Now, however, something didn't look right. I don't remember any rocks in the middle of the channel there, he thought.

Gradually, as he closed on the sight Ng and his copilot could make out the form of a large crab boat–sized vessel. Waves were crashing over her bow as she pounded her way at top speed through the tall and wintery seas.

The vessel turned out to be the 121-foot Marine Vessel

37

Wolstadt, an Alaska Department of Fish and Wildlife boat. She was covered in ice and captained by none other than an Alaska state trooper known from Kodiak to Dutch Harbor as Lt. Bob Lockman.

Ng was astonished to see him there. And as he flew by overhead, he radioed down to Lockman on the VHF channel. Ng's voice was filled with awe and admiration.

"How you doing down there? Are you all right? Heck of a day to be taking a joyride! Boy, you look like you're having a terrible time."

"Well, yah, it's pretty rough," shot back Lockman. "We were just wondering what anybody in his right mind would be doing flying on a day like this? What *are* you boys doing out here flying in this weather?"

"We're headed for the *Mia Dawn!*" replied Ng.

"Well, I kind of figured as much. That's where we're headed, too," countered Lockman. "Figured we might be able to help 'em out. I guess that makes us both crazy."

At about Kupreanof Point, Ng and his crew broke out to great visibility and a high blue ceiling. Then the tall white peaks of the Aleutian Range came into view with spectacular clarity. "To the north we could see the williwaws and whiteout areas building at the head of each pass," recalls Ng. "A weather system, a high, had built up on the Bering Sea side of the Aleutian Range, and the winds were really whipping up."

What was happening right before his eyes was that untold tons of powdered snow were being stripped from the ice fields high in the mountains and blown out of the steep mountain passes as if caught in a wind tunnel. The howling vacuum tore the wave-tops off the ocean waves and sucked it up in its vortex, creating a sticky, blinding layer of snow, ice, and freezing spray. This "solid wall of solid white" was capable of sticking to, weighting down, and sinking any vessel caught in its path.

As Ng watched, the refrigerated vacuum of arctic air came howling down out of the ten-thousand-foot-high passes, blowing magnificent breathlike blasts of powdered snow that stretched for twenty miles and more offshore, out across the heaving face of the Gulf of Alaska sea.

As Ng and his embattled crew of six neared the rescue site, an ever-increasing barrage of ice-laden arctic winds drove into them.

Ng put out a radio call to any boats that might be near. "Hello to all listeners! This is the pilot of a U.S. Coast Guard helicopter flying near Cold Bay! We are searching for the fishing vessel *Mia Dawn.* Is there anyone in the area who has any information that could help me locate the *Mia Dawn?*"

Skipper Paul Vines, aboard the fishing vessel *Bob's Boat,* answered the call. "The *Mia Dawn* is somewhere west of us!" he said.

When he saw them idling just outside the impenetrable wall of howling white, Ng's copilot, Lt. Com. Larry Cheek, knowing full well the fierce loyalty among Alaskan fishermen, said aloud, "If they know their comrade is in trouble inside that whiteout there, why aren't they going in there themselves?"

Not long before though, the *Bob's Boat* had attempted to push its way to the *Mia Dawn,* but the wind had proved to be too powerful, and he was driven back. "They couldn't get in there," recalls Ng. "They were blown right back out of the bay!"

The winds were so ferocious, he soon learned, that the few crab boats in the area had taken shelter from them by dropping anchor in the lee of a large mountain. Even so, they were forced to keep their engines running and propellers engaged, for as quickly as the fierce Alaskan winds found them, they were swept one by one back out of the bay, their huge steel anchors dragging like tin cans behind a newlywed's car.

Every few minutes now, those who called to advise Ng were forced to "leave their radio" periodically in order to climb up on their wheelhouses and break the ice off their radar screens and antennas.

Then the skipper aboard the *Bob's Boat* radioed Ng back. He'd had contact with the *Mia Dawn*. "I can still hear him. He's over to the west of us!" The skipper couldn't be certain exactly how far away the terrified crewmen were. But conditions, he assured Ng, were "Unbelievable!" The wind-meter gauge mounted on top of his 110-foot crab boat had just been broken off by a gust of wind that had topped 115 miles per hour.

For several minutes then, Jimmy Ng hesitated, hovering over the water. In their present position, the atmosphere whipping around them was cold and clear. But directly in their path now stood what looked to be an opaque wall of arctic wind and blowing snow several thousand feet high, and tens of miles long, all hurtling out to sea.

It was a strange and frightening anomaly of nature. And as Ng flew ever nearer to the fearsome formation, it rose out of sight overhead. The thought of entering such a thing filled Ng with foreboding.

It was then that Ng picked up radio contact with the skipper of the *Mia Dawn*. The connection fluctuated between tenuous and nonexistent. But from the short, static-filled exchange, Ng could make out that the boat had run aground on some rocks, that it was being pounded by the wind and seas, and that she was icing badly. Her crew had managed to break out their life raft, but they had inflated it only halfway when the *Mia Dawn* rolled over onto the raft, pinning it beneath her hull.

But Jimmy Ng and his copilot had major mechanical problems of their own. For the last several hundred miles, they'd been flying by "visual navigation" only, since both the helicopter's radar and loran systems still refused to function.

In an effort both to hear and clarify the radio calls coming from the crew of the *Mia Dawn,* Lt. Com. Cheek disabled the squelch on his radio. In theory, this would allow more of the faint, erratic radio waves coming from the *Mia Dawn* "to be accepted." The technique worked: The once-deafening static that had been breaking up their calls now came through much more clearly and consistently—paradoxically, making any incoming signals clearer, as well.

Cheek decided to call the captain of the *Mia Dawn,* asking him to give both a "short and a long count" over their radio. The purpose of this one-to-ten count was to get them to stay on their transmitter long enough to allow Ng to use the copter's direction finder to locate the stranded boat.

"Hello, *Mia Dawn!* Hello, *Mia Dawn!* This is a U.S. Coast Guard helicopter!" announced Jimmy Ng.

But when the voice from aboard the *Mia Dawn* replied, it was with a frantic cry: "We definitely *are* in need of assistance! We're not going to make it without help! We can't hang on much longer! We need you here *now!*"

Saving lives was everything a Coast Guardsman was about. Ng knew he had to make a decision. As the danger level increased, his adrenaline started flowing. "And when it flows," he recalled years later, "your mind gets very sharp. You start looking at everything in fine detail. You want to be as precise as you can."

Captain Ng briefed his crew on the facts of the situation.

Speaking over the radio, he informed them that the radar and loran systems had both died and that the aircraft was getting low on fuel, but he said that they were getting close to the vessel in distress. Finally, he added that the crew of the *Mia Dawn* would probably not survive if they were not rescued immediately. "It's now or never," he said finally.

Ng and Cheek knew that they "had to pinpoint their position as best they could "before" they went in. Taking that kind of precaution prior to entering the frightening gray-white vacuum rising directly in front of them would be essential to their success and their survival.

Lt. Com. Larry Cheek continued to listen to the static-charged voice of one of the frantic *Mia Dawn* fishermen. He was trying to pinpoint the vessel's location, doing his best to detect volume changes in the crewman's voice while struggling at the same time to maneuver the wind-battered chopper. Ng hovered just outside the main body of the howling vacuum, and, at his copilot's direction, pivoted the helicopter one way and then the other in an effort to zero in on the *Mia Dawn*'s location.

"*Mia Dawn*," radioed Cheek, "I need you to continue to give me a long count! I need you to count from one to ten again slowly while continuing to key your transmitter. Then count backward from one to ten while again continuing to key your transmitter."

As the unidentified voice aboard the *Mia Dawn* continued to count aloud, Cheek put the instrument into a direction-finding mode. The needle leapt to life, pivoted on its axis, and froze at an entirely new heading. To double-check its accuracy, Cheek suddenly shut the machine off. The needle fell lifeless. When he switched the direction finder back on, the needle locked right back onto the very same heading.

Indeed, the needle appeared to be tracking the voice, and it offered a consistant heading. Ng was hovering at about sixty feet above the water when he radioed his crew. "Let's go ahead and take a look and see what it's like in there," he said. And so, without radar with which to see, or a loran computer with which to navigate, Jimmy Ng inched ahead toward the incredible white wall awaiting him.

With their pleas for help echoing in his ears, Ng brought his altitude down in the hope that, once inside the strange phenomena, he might be able to keep visual contact with the ocean surface. And with this in mind, he began "hovering in" to locate the foundering hull of the *Mia Dawn*, lying unseen somewhere off in the howling mist ahead.

Facing upwind, Ng taxied sideways into the mass of snow and freezing spray standing two-hundred stories tall in his path.

Instantly, the water below Ng disappeared. The belting force of the wind struck like a giant fist and the chopper spun out of control.

Ng remembers being "thrown around as if we were nothing!" He battled to bring the chopper back, but the wind continued to manhandle it. Ng knew immediately that he "had made a big mistake." The struggling aircraft went into a free fall, recovering only a few yards from the water. In the next moment, the chopper soared more than one hundred feet into the dimensionless heavens above — only to plunge again toward the 39°F. waters of the Gulf of Alaska.

Again, the Coast Guard helicopter, with six tense crewmen aboard, dipped perilously close to the ocean's surface before Ng could halt her descent and "bring her nose back up."

Ng knew that the rotor blades of a helicopter that has landed on a flat, calm sea will clear the water by only a few feet.

But the crew of a chopper forced down under such horrendous conditions of wind, ice, and sea would face catastrophe. The helicopter's blades, all five of them, would strike the water first. Plowing into it at 8,000 rpms, they would explode into pieces, shattering like icicles striking concrete upon impact. Then the chopper would auger in—crash—and in such high seas, she would roll over, most likely trapping whoever still remained inside the aircraft.

Ng's windshield now began picking up a coating of ice. The men aboard the helicopter knew that if they went down "that far from everybody, there would be nobody to rescue them," recalls Ng. As the craft leapt, bolted, and free-fell, those aboard knew there was nothing to be done but "ride it out." "We were all scared to death," recalls Ng.

As Ng flew on, Cheek listened intently to the long count coming from a crewman aboard the *Mia Dawn,* at the same time keeping his eye glued to the minute arrowlike arm oscillating like a compass needle inside the tiny two-inch-wide DF gauge mounted on the instrument panel in front of him.

No matter how badly the helicopter rocked, no matter how high his own personal "pucker factor" rose, it was his duty to guide and assist Ng, his commanding officer. And while backing Ng up at the controls, Cheek fed him a steady flow of course corrections as the men flew blindly on.

Most encounters with severe turbulence are of short and frenetic duration. But this encounter was momentous and unrelenting and tossed the bulky form of the H-3 "Huey" helicopter through the air in a ceaseless series of harsh, battering blows.

Charts and tools—everything not tied down—were tossed helter-skelter inside the cargo space of the plane. In the cockpit, pencils and maps went flying.

"At Disneyland, it would have been an 'E' Ride," recalls Ng.

Dr. Martin Nimeroff, a hypothermia expert, was one of the crew on board during that flight. He was being banged around violently as the copter battled the storm, and found that there was only one way he could keep from being slammed from side to side—he had to make an X form out of his body. Spreading his legs wide on either side of the floor and bracing his hands equally wide against the ceiling overhead, he tried to ride it out.

The wind was blowing so hard, Ng found himself unable to turn the helicopter around, because the tail rudder—the small propeller mounted on the tail and used to direct and steer the chopper—was completely overwhelmed by the typhoon force of the winds. He was forced to keep the helicopter faced into the wind the entire time. "Fly five or ten degrees to either side and you would lose control," recalls Ng.

All this time, Lt. Bill Gottschalk, a respected and highly experienced pilot, and his co-pilot, Lt. Fred Darvil, were clinging to the controls of the 130,000-pound U.S. Coast Guard C-130 SAR plane circling high overhead.

"It's clear above! It's clear above!" Gottschalk radioed from above.

"Well, it's not clear down here!" shot back Ng.

Gottschalk knew that Ng's helicopter had lost most of its essential navigational instruments. Now, hearing that Ng was in trouble, Gottschalk decided to descend in an attempt to secure visual contact with the U.S. Coast Guard helicopter crew struggling several thousand feet below him.

But he had hardly begun the maneuver when he ran

into the most violent air currents he had ever encountered in his life. "The turbulence was so severe," recalls Gottschalk, "I couldn't even attempt to penetrate it." Nor could he even imagine that Ng and Cheek were "down there hovering" in such incredible winds in the cumbersome hulk of an H-3 helicopter, let alone attempting to hoist people aboard.

In more than a decade of flying SAR missions for the U.S. Coast Guard in Alaska, Ng had "never flown in anything close to the turbulence" he experienced that day. "Big bombs, williwaws, of air, would come rolling down out of the mountains and hit the water and just explode!" he recalls.

"Pow! Pow! Pow!" sounded the williwaw gusts as they slammed into the sides of the chopper.

Each time another jolt of williwaw wind struck, Ng would fight to regain and then maintain control of his craft.

Once, as Ng glanced down at his instrument panel, he noticed that the wind-speed indicator was pinned over on one hundred knots—about 112 miles per hour. Then, looking down, he caught a brief glimpse of the foam-streaked surface of the ocean beneath him. He was astonished by what he saw. His ground speed was "absolutely zero." He wasn't even moving.

As each jolt of wind struck, Ng felt himself lose control of the helicopter for a few terrifying moments. "It bounced us around so bad," he recalls, "at those times we were just along for the ride. We were scared. There was no doubt about it. But then we would hear those guys aboard the *Mia Dawn*—who were dying."

Suddenly, Ng realized that he had ventured well beyond the capabilities of both himself and his helicopter. The reality of his predicament closed suddenly on him. Ng's mind raced. Shouldn't be here! he thought, chastising himself. Could crash. Could very well die. What can I do to get us out of this?

Then, navigating blindly through the snow mist of the howling ice storm, with the lives of his entire crew in imminent peril, Ng applied everything he knew in a fly-by-the-seat-of-the-pants attempt to survive and regain control of the helicopter.

Ng knew he had no choice now. He flew through the blinding snow mist, navigating on pure instinct. We shouldn't be in here, he thought again. This is beyond our capabilities. Ng, his helicopter, and his entire crew were perilously close to being literally "blown out of the sky." And with no way to see or locate the pleading crew of the *Mia Dawn,* Ng, in his heart, gave up the search. Under the present circumstances, the best he and his copilot could do would be to bring themselves and their entire crew out of the storm alive.

It had been the hardest decision of his life. In order to save his own flight crew, he would have to abandon the search and ignore the pleading cries from the men aboard the *Mia Dawn* — effectively abandoning them to God, the wind, and the insufferable cold. It was his duty to bring his own crew out alive. Now he weighed whether to escape by continuing to travel the way he'd been going, or to try and hover back out the way they'd come.

"Well, it can't be any worse ahead than what we just came through," Ng radioed to his copilot. "Let's just keep going west." Copilot Larry Cheek agreed.

Just then, a frantic radio call blurted over their radio headsets. It was a crewman aboard the *Mia Dawn. "We can hear you! We can hear you!"* a terrified voice yelled. The fisherman's words riveted Ng and brought him "back into the rescue."

"Mia Dawn! Mia Dawn!" Ng called back. "In which direction do you hear us? Where are we from you?"

"You're right above us!" came the incredible reply.

Ng rushed to radio his own crew. "Hey, everybody, look around now and see if you can spot this boat!"

The radioman sitting beside a window on the left side of the helicopter spotted them first.

"I see them!" he radioed suddenly. "I see them! They're right underneath us!"

In the howling assault of wind and ice, Jimmy Ng slid backward for a closer look at the *Mia Dawn.* For a few moments, as the blowing sea spray and ice flew past, Ng couldn't really tell whether there was "anything other than ice there." The mysterious image could well have been just a chunk of glacier ice drifting off into the blinding haze.

"It might have been the profile of a boat—or something you just wanted to see," he said later.

Then there was a millisecond break in the whiteout, and for the first time Ng was able to take in what he described as the "ice figurine" form of the *Mia Dawn.* The vessel was lying on her side, at about a fifty-degree angle. The *Mia Dawn*'s life raft, now only partially inflated, lay pinned beneath the tilting slope of the wheelhouse. Every inch of vessel above the ocean surface was encased in more than a foot of freezerlike sea ice.

But something more was eating at Ng. He'd been told that the crew members of the *Mia Dawn* were equipped with survival suits. Now he discovered that they were dressed in everyday work clothes. "Not a one wore a survival suit," recalls Ng. Unless Ng and his crew could hoist them aboard, he knew that the moment the *Mia Dawn* went under and her crew members spilled overboard, there would be no way to gather their scattered bodies, and the quick and deadly procession of hypothermia would begin. To freeze to death would take only minutes. The crew of the *Mia Dawn* wouldn't have a chance.

The vision before him brought to mind the 1974 photos of the *John & Olaf,* found aground and abandoned under exactly the same circumstances in a legendary ice storm in another Alaskan mainland bay not far from that very spot.

Rescuers had tried for several days to reach the Maydaying crew. But faced with williwaw gusts that reportedly topped 150 miles per hour, rescue ships and aircraft alike were driven back. For several days and nights, no one could reach the foundering vessel. When rescue crews finally did reach the *John & Olaf,* she was sitting hard aground on a reef. The entire vessel was wrapped in a specter of frozen spray. Whole feet of solid ice coated every exposed inch of her superstructure. She more closely resembled the bowels of a freezer than the exterior of a fishing vessel. But the ship's life raft and crew were nowhere to be found. Inside, a cup of coffee sat undisturbed on the galley table. No trace of any of her four crewmen was ever found.

Whenever the speed of storm winds blowing over a space of open seawater exceeds sixty miles per hour, the wind begins to tear the tops off the ocean waves and sends that wet ocean matter hurtling horizontally across the face of the sea. In the battering, one-hundred-knot blasts of wind in which Ng now found himself caught up, this wall of spray rose several thousand feet into the sky.

Ng rightly worried that the freezing salt water pummeling them would soon impair the aircraft's engines. Power loss, he knew, would be the first warning sign—followed by total engine failure.

Perhaps no helicopter pilot in American peacetime history had ever found himself caught in a more impossible predicament. Buffeted by headwinds in excess of one hundred miles per hour, wandering blindly through a whiteout in search of an unseen vessel, with virtually no navigational equipment functioning on board and no more than fifty feet of visibility in any direction, with Siberian-born winds packing subzero temperatures, and a spray that continued to coat the windshield and skin of his struggling, straining helicopter in ever-thickening layers of leaden ice, and, finally, with little more than thirty minutes of precious fuel remaining, Ng made another judgement call. He would do battle with the stupendous gusts of wind and the growing burden of ice and attempt to hover in place long enough to rescue the doomed crew of the *Mia Dawn.*

"Okay," he said, speaking in a steady voice over the radio to his crew. "Let's try to hoist them."

As the flight mechanic aboard the aircraft, Michael Barnes sat on the right side of the chopper, next to the sliding door. Barnes slid the cabin door open and was faced with a howling vacuum of wind and ice screaming past only inches away. He rushed to hook up the body basket and get it ready to go. As he did, buffeting winds slapped and rocked the chopper. He took in the loud whine of the gearbox beside him. It was under a tremendous load as the chopper battled to remain upright and airborne.

To Michael Barnes, it sounded as if their chopper was in the very act of "coming apart."

"Sir, are you sure we want to do this?" Barnes radioed his skipper.

"Well, we're here," shot back Ng. "We have to do all we can now."

Though rescue conditions in Alaskan waters are often severe, the heavy steel body basket usually proves weighty enough to allow Coast Guardsmen to lower it alone to the survivors below. But that was not the case this day.

When Barnes pushed the mummylike form of the U.S. Coast Guard's steel body basket out the side door and began lowering it, the basket immediately sailed about wildly like a Frisbee in the incredible turbulence.

"I can't do it! I can't do it!" radioed Michael Barnes to Ng. "I can't get the body basket to go down. The wind's blowing too hard!"

Ng replied immediately "Yes you *can* do it!" he said. "And you are going to do it! Figure a way to get that thing down there!"

"We're going to lower a line down to you," radioed Ng to the *Mia Dawn* below. "Once you get ahold of it, tie the line to your boat. Then pull the basket down to you."

Buffeted by the tremendous velocity of the wind, Michael Barnes pulled the body basket back inside. Then he secured a length of tag line, tied several small sandbags to the end of it, and tossed the weighted end out the door to the men below. Instantly, the wild velocity of the wind and ice matter blew the line horizontally through the air, very nearly tangling it in the tail rotor blade. If that happened, Barnes knew, it would almost certainly bring them down.

"Sir, our tag line is blowing aft. It looks like it could go into the tail prop. The winds are just too strong. I've got to pull it back in."

"Then bring it back in," replied Ng. "We don't want to crash with them."

Barnes jumped to the task. He jerked the tag line away from the tail prop blades and hauled it back on board in short, decisive, hand over hand movements.

Again and again, the helicopter pitched savagely through the air, as Ng worked to maintain a hover some fifty feet above the *Mia Dawn*. All the while, the chopper blade rotor gears were maxing out in Michael Barnes's ears. The unsettling sound was both eerie and deafening.

Then, tethered to the interior of the chopper by a body harness, Barnes leaned out the sliding door into the elements. All he could think of was "getting the job done and getting out of there."

He could see the three crewmen scrambling across the frozen form of the *Mia Dawn* below. She was lying on her side. Her back deck was packed tightly with the large domino-shaped forms of crab pots. The men were trying to clear a place for the Coast Guard body basket to land. And as they did, waves crashed in repeatedly against the side of the foundering vessel, launching bone-chilling geysers of freezing spray over the boat and the men.

When Michael Barnes pulled the trail line back inside, he was stunned once again at the extraordinary velocity of the howling vacuum of wind and ice blowing past him only an arm's length away. His mind raced. Suddenly, it came to him. The sea anchor! There was a forty-pound chunk of sea anchor that came inside the life-raft package stored on board the helicopter. He raced to get it.

Barnes moved back to the sliding door and the winch controls, where the wind and ice and snow were blowing fiercely as ever. As the helicopter staggered back and forth through the gusting stratosphere, he once again spotted the half-hidden outline of the *Mia Dawn* and the hazy forms of crewmen scampering across her. The vessel was now ahead of them. Immediately, he began "conning" the chopper forward.

"Sir, move forward thirty [feet]," he radioed Ng. Almost as quickly as he had requested it, the helicopter began inching forward into the battering winds. As the chopper pushed slowly ahead, Barnes fought to control the emotion in his voice.

"Forward twenty!" he radioed. Then, shortly, he transmitted the instruction "Forward ten!" Suddenly, a powerful jolt of williwaw wind drove into them. The helicopter pitched sideways and plunged towards the sea. Ng rushed to compensate. The whining tone of the gearbox rose feverishly; it sounded like an incoming shell just prior to impact.

Ng was able to bring the nose up only yards shy of colliding with the ocean surface. It would take Jimmy Ng another ten minutes to battle back into position.

When Ng finally maneuvered the chopper into what Barnes believed was the optimum position, he announced simply, "Hold!"

Barnes could feel his captain struggling to keep the wind-battered chopper hovering in place. Racing to complete the task at hand, Barnes slid the weighted end of the trail line over to the door and pushed the anchor out over the side. Even with the additional weight, it did not surprise Barnes when the trail line blew sharply aft. In one motion, he began playing out the line in an effort to feed it down to the frantic crewmen below.

"Trail line being lowered, sir," he radioed. Then he said, "Trail line going onto the deck, sir."

The men on board the *Mia Dawn* grabbed the anchor and pulled the trail line down hand over hand. The body basket went next. But the moment the steel basket arrived on deck, it slid down cross the icy incline and became lodged among several of the crab pots.

From directly overhead, Barnes could see the problem. But the crew of the *Mia Dawn* seemed to ignore it. Worst of all, instead of freeing the basket, they began loading personal belongings into it.

"Hey, they're throwing their stuff in! I want *them,* not their gear!" he radioed Ng.

It wasn't so much the added weight that troubled Barnes as the amount of time they were using up. No one could predict how long Ng would be able to hover in place—or even remain in the air, for that matter. Worse yet, the helicopter was running critically low on fuel. Only minutes of the precious liquid remained. If it did run out, the chopper would literally fall out of the sky. And then no one would be going home.

But Michael Barnes found it impossible to communicate that to the crew below. Several of the crewmen continued to load radioes, clothing, and personal items into the rescue basket.

When one of the shaggy, bearded crewmen on the *Mia Dawn* finally climbed into the basket, Barnes grabbed the hydraulic winch lever mounted alongside the chopper's side door and pulled. The winch spool jumped to life and the steel cable jerked taut.

Then, racing against time, Barnes "pinned it, maxing out" the winch's capacity as he lifted the crewman and an unwelcome assortment of gear skyward to the chopper. Moving at approximately two hundred feet per minute, the basket and its cargo rose quickly.

As is common under severe conditions, as soon as the basket lifted off from the deck of the *Mia Dawn,* the fierce storm winds caught the broad "sail area" of the loaded basket and sent it spinning wildly through space. "It was quite a ride coming up," recalls Barnes.

As the basket neared to within ten feet of the chopper's side door, the winch speed automatically slowed to approximately fifty feet per minute. This was a built-in safety feature, designed to keep those inside the basket from being crushed on the bottom of the hovering chopper by an overly excited winch operator.

The crew member rose to just below the chopper's side door. Then Barnes extended his leg, and, using it as a bumper, ended the basket's dizzying orbit.

As Barnes pulled the grateful crewman aboard, the young man tried to speak. But with his helmet, the incessant roar of the wind, and the chopper's engine all muffling the sound, the man couldn't make himself heard.

Barnes said he was "more than a little pissed" when he got the man aboard. As soon as the crewman was out of the basket, Barnes began tossing the fisherman's personal belongings out the sidedoor and into the sea with a furious indifference.

Worried sick at the diminishing level of fuel on board, Barnes was infuriated when, having lowered the basket for the second of the three crewmen, he saw the men below loading a boom box and even more clothes into the rescue basket.

Through each journey, Barnes continued to brief his skipper as each phase of the rescue occurred. "Sir, your altitude is good. Your position is good. One man is in. . . . Taking the basket down for the next crewman. Lowering basket. Basket's halfway down. Basket's on the deck. Man getting in the basket, sir. . . . Preparing to take the load. Taking the load. Basket is clear of the vessel. Bringing him up. . . . Man is coming aboard. Man is out of basket and inside the cabin. Lowering basket for next survivor."

When the third and last crewman had been pulled aboard, Barnes quickly untied the line and tossed it out the side door.

Outside, the howling, gray-white vortex of blowing wind, spray, and snow continued to race past. Occasionally, as the helicopter leapt and fell beneath them, Barnes caught brief flashes of the sea itself. The seawater lay dark and green, broken only by thin streaks of foam and the brilliant white flashes of storm waves breaking.

"Hoist secured. Door closed," Michael Barnes radioed Ng finally. "Ready for forward flight!"

Now, with the three fishermen hoisted safely on board, Jimmy Ng faced his old dilemma: how to fly out *and* survive the passage. Ng knew that his options were few. They could attempt to return by the same route that they had come. But Ng knew there was a good chance they "would get beat up real bad and maybe crash."

He knew for certain that he could not move forward very far. For one startling moment, as Ng and his crew were struggling to hoist the *Mia Dawn* crew aboard, the blinding curtain of williwaw gusts had momentarily parted, revealing a frightening vision. There, only a few hundred feet in front of him, stood the sheer rock face of a mountain. The mountain rose from the sea itself and ranged high above them, its sides marked with intermittent streaks of black rock and white ice.

"We couldn't go ahead because of the mountains," recalls

Ng. "And we couldn't fly sideways in either direction because of the incredible velocity of the williwaw winds."

They could try and back down. But the turbulence was so great that it would probably have dropped them "out of transitional lift and down into the water," where the tall seas would quickly roll them over.

Again they even considered continuing west in the same direction that they had been going. But things were getting worse, not better. Ng and his copilot talked it over. "Well," said Ng finally, "what I'm going to do is go full collective— I'm going to pull our nose up ten degrees and just try to go straight up." Another term for *full collective* is *maximum power pull;* Jimmy Ng was going to try to fly his way up and out of the storm.

With the cliffs rising unseen off in the whiteout ahead of them, Ng told his copilot, "We have to keep the nose of the chopper up or we'll fly right into the mountain."

And with that, he began his ascent. Ng lowered the helicopter's nose slightly to gain some forward ground speed, then attempted to turn to a course of 170 degrees. "We pulled power and went to one hundred knots of airspeed almost instantly," he recalls. Even with the power maxed out, the turbulence leveled the aircraft off at only two hundred feet for a few scary moments, and it refused to rise higher.

Battling back, Ng eventually climbed past the three-hundred-foot level. It was then, as he was accelerating up through the formless, battering white mass, that he felt the heaviness of exhaustion close suddenly on him. Three helicopter rescues in a single day, including some seven hours of almost-continuous high-stress flying time, represented an astonishing workday for even the most experienced of pilots. Momentarily overwhelmed, Ng now began to experience a touch of vertigo.

It was Ng's alert backup pilot, Larry Cheek, who noticed it first. He "jumped in and probably saved my and everybody else's butt, as well," recalls Ng.

"Watch your nose!" Cheek warned Ng. "You're dropping your nose."

"Yes, I see it," replied the fatigued Ng. But the nose continued to drop and the helicopter accelerated forward.

Ng was "really tired," and through the fatigue and accompanying vertigo, he told himself to "pull the nose of the helicopter back up." But he was having trouble "getting it right." The moment he sensed his skills were slipping, he relinquished control of the helicopter.

"Take the controls!" he said to his copilot.

"Okay, I've got it!" shot back Cheek. He grabbed the controls, brought the chopper's nose up, and continued the climb.

At an altitude of approximately 2,500 feet, the helicopter suddenly broke free of the blinding mass, rising up into a crystal-clear world filled with the stark, mountainous beauty of the Aleutian Range standing upright and unforgiving against a heavenly blue sky.

"Thank God!" said Michael Barnes, looking down on the blinding williwaw below.

With the chopper engine running almost on vapors, they flew directly to the nearby Aleut fishing village of Sand Point. They landed at the tiny island airport with only a few minutes' worth of fuel remaining.

After hours filled with the high-risk stress of flight and rescue, the chopper had finally landed safely. When its engine was shut down and its chopper blades finally twirled to a stop, the utter silence that followed was a peaceful respite.

"At sea, the crewmen on board the *Mia Dawn* had been very glad to see us," recalls Ng. Yet now their fatigue left them almost speechless.

"That's one heck of a machine you've got there," offered one of the fisherman. "I didn't know a helicopter could do that."

Jimmy Ng, Larry Cheek, and the rest of the exhausted crew left the helicopter at Sand Point to be checked over by mechanics and caught a ride back to Kodiak on a C-130. "It [the H-3] had been rung out pretty good, and we wanted her thoroughly checked over," recalls Ng.

Lt. Bill Gottschalk, the pilot of the C-130 that had been circling them during the rescue, was waiting for them on the ground at Sand Point. "Do you know what you said to me after you landed at Sand Point?" he asked Ng several weeks later.

"No," replied Ng, unable to remember.

"Absolutely nothing," Gottschalk replied. "You didn't say a thing. None of you said so much as a word all the way back to Kodiak."

Ng knew why. "We'd all been too close to death," he said.

Captain Jimmy Ng, Lt. Com. Larry Cheek, and the entire crew went on to receive the Distinguished Flying Cross for their heroic efforts during the rescue of the crew of the fishing vessel *Mia Dawn.* This is the highest peacetime medal a serviceman in the United States can be awarded.

"For being in the wrong place at the right time," as Ng describes it, he and his crew also received national and international awards for "Best Military Pilot and Crew of 1985." Furthermore, Ng was chosen by candidates from the Navy, Marine Corps, and Naval Reserve to receive the Helicopter Aviation Award for valor, an annual honor presented by the Association of Naval Aviation.

Finally, for the heroic rescue of the crew of the fishing vessel *Mia Dawn* during that winter of 1985, Ng and his men received the greatest honor of all: They were singled out for having performed the most exceptional rescue of anyone among all branches of the U.S. Armed Forces serving anywhere in the world.

Journey of No Return

In my first book, Working on the Edge, *I told the tragic story of the crew of the F/V* Saint Patrick *as seen through the eyes of Wallace Thomas. Thomas was the lone survivor of a group of ten crewmen who, in their dark panic, tied themselves together with ropeline and leaped overboard off the stern of the F/V* Saint Patrick.
The first crewmen to abandon ship, however, were Bob Kidd and Clifford "Doc" Steigel. They were thirteen miles off the rugged coast of Kodiak Island when, on that stormy November night, they decided to leap overboard from the bow of the Saint Patrick. *Only Bob Kidd would survive the incredible ordeal ahead.*
Never before told, this is his story.

During the winter of 1981, the 155-foot ship *Saint Patrick* idled along the Kodiak waterfront and out onto the wilderness waters of the Gulf of Alaska in search of scallop. That night, a brutal storm packing ninety-mile-per-hour winds and forty-foot seas would surprise them. The journey would be filled with bad luck, bad decisions, negligence, and ineptitude. And it would serve to rewrite U.S. maritime law. For the majority of the twelve crewmen on board, it would prove to be a doomed voyage: a voyage of no return.

It was near midnight on November twenty-ninth. The winds outside were accelerating past eighty knots and the storm was building fast when, rousted from his bunk by several terrified young deckhands, thirty-one-year-old deck boss Robert Kidd climbed into the wheelhouse of the fishing vessel *Saint Patrick* to check things out. He could hardly believe what he saw. The ship was drifting sideways—entirely out of gear—through the large and ever-building seas. Rain and ocean spray were drenching the outsides of the windows in melting waves, blinding all vision.

On the starboard side of the wheelhouse, a young and inexperienced crewman was standing at the helm, fighting to gain control of the ship. The port-side window had been rolled down and another equally unproven crewman was hanging out the window, shouting directions. "Go left! Go more left!" he yelled, a neophyte's plea for a portside maneuver. The skipper was nowhere to be found.

"Hey!" yelled Bob Kidd to the one doing the steering. "Are you going to get this thing turned into the seas, or what are you going to do?" Kidd knew instinctively what they ought to have been doing: They needed to "quarter" into the waves, a strategy designed to lengthen the time between the tall wave crests; it was a way of breaking up their tight and deadly cadence and allowing a vessel time to recover before the next wave struck.

"I'm just trying to keep it from sinking!" the terrified young crewman shouted back. "I'm trying to keep it afloat! She's going to sink if we can't get her in line!"

"Well, why the hell aren't you using the hydraulics?" shot back Bob Kidd, pointing to the controls in front of him.

The crewman could only shake his head. He didn't know anything about them.

Kidd approached quickly and instructed him. "Here! Just hit it hard to port and put her into those seas!"

The young man obeyed and the ship immediately began to swing around.

Kidd went below then. Once below, however, he felt the waves again begin to explode against the port side of the *Saint Patrick*. They were driving directly into the hull; it was immediately clear to Kidd that the pilot had lost control of the ship and was once again drifting sideways to the huge storm waves.

Also aboard the ship that night was the tall, muscular crewmate known as Doc Steigel. Doc and the shorter but stockier Kidd were the most experienced men on board ship. They knew that the *Saint Patrick* should have been moving ahead bow-first, directly into the mounting seas. When no change came, Doc appeared at Kidd's stateroom door.

"Bob, I'm scared," said Doc. "These guys don't know what they're doing! Something's wrong—you've got to go up there and take over."

"I can't do that!" shot back Kidd. "The captain is still the captain. I could go to jail for doing something like that!"

"But Bob, the skipper is just lying in his room! It ain't right!" Kidd could see the fear in Doc's eyes. Both men had a healthy respect for the force of the sea. And what was taking place up in the wheelhouse at that very moment ran in direct contradiction to everything they knew to be true about navigating in high seas.

Though William Green was a friendly and at times jovial skipper, Kidd had never trusted in his competence as such. Once, Kidd and another crewman had been up in the wheelhouse, watching Green try to navigate out of the harbor at the Kenai Peninsula town of Homer, when the other crewman began to laugh aloud. Green's eyes, it seems, had been glued to the radar—while the *Saint Patrick* was moving ahead on a collision course directly toward shore.

"He didn't even know it," Kidd recalled. "Scott told Green, 'You know, you're heading for land,' and Green replied, 'Yes, I know, I know,' and, looking up and seeing land, he immediately corrected his course."

So now Kidd went searching for his skipper, whom he found practically cowering in his bunk. Infuriated, Kidd marched into his room and confronted him.

"What the hell is going on?" he demanded. How could you leave a completely inexperienced crewman at the helm in such a storm?"

The skipper was openly distressed; tears welled in the man's fear-filled eyes. Rising to return to the wheelhouse, the skipper complained aloud, "It was only supposed to blow fifty miles per hour!"

Suddenly, a jolting shudder reverberated through the entire length of the steel ship. The *Saint Patrick* had been "stopped dead" in the water. To Kidd, it felt as though they'd run aground on solid rock.

As he was racing to get his survival suit, Kidd ran into Doc Steigel hurrying toward the wheelhouse stairway.

"What happened?" Steigel asked. But Kidd had no answer. He suggested that the two of them make their way outside to survey the situation.

Steigel agreed, but not without first putting on his survival suit. "If we get washed off that deck without them, we won't have a chance!" he warned Kidd.

As Kidd and Steigel ran through the galley, the floor shifted beneath their feet again, the microwave oven came out of the wall and smashed on the floor, and "things started flying."

Doc Steigel and Bob Kidd raced to pull on their survival suits, then worked their way outside and down onto the wave-washed deck. The wind was blowing a steady eighty knots (approximately ninety miles per hour) now, and as it raced through the night it howled against the superstructure and cable rigging of the *Saint Patrick,* filling the close canopy of darkness with stinging particles of ocean spray and washing her deck in thunderous episodes of collapsing sea.

As they investigated for damage, the engineer stuck his head out the wheelhouse window.

"Hey! It's no problem!" he screamed down to them. "We just took a bad wave on the port bow!"

Yet incredibly, Kidd noted that the ship was once again drifting sideways to the lumbering approach of the stormy seas. Perhaps it was the steering problem giving the skipper fits again. Regardless, it was a dangerous situation.

"I'm not going back in," shouted Kidd. "I'm staying right out here on deck, and I'm leaving my survival suit on. These guys got it side-to in the waves! These seas may very well flip us over! And if that happens I want to be out here!" Doc agreed. Being trapped inside a sinking ship meant certain death. Then, from behind them and far overhead, came the building roar of what sounded to Kidd like "a hundred trains." He later remembered reeling around just in time to see a wall of ocean water the likes of which he had not seen in his entire life. "I looked up. And I just kept looking up and up and up." And then, out of the coal black depth of night, Kidd's dilated eyes took in the terrifying approach of a mountainous wave. It was wild and tumultuous, and topped with a curler that gleamed fluorescent white in the reflective glow of the ship's mast lights.

It was the largest curler mounted high atop the largest wave Kidd had ever seen. "If it wasn't sixty feet, it was nothing!" recalls Kidd. And it towered over the ship's twenty-five-foot-high smokestack, and, of course, the tiny forms of the two crewmen on the wooden deck below.

Dwarfed by the awesome dimensions of the wave, the two men were caught directly in its path. Kidd was certain he was a dead man. Nothing could oppose such a force. Then everything around him grew strangely quiet; even the roar of the mountainous wave overhead seemed to hush and take pause. For Bob Kidd, it was as if he was lost in a sensory vacuum, and in that frozen moment he could no longer feel the bitter cold of the spray, nor the icy November winds that drove it; nor was he conscious of the heart-stopping fear that filled him. The sudden and terrifying roar of the wave powering toward him had just as suddenly grown silent.

Mesmerized by the gigantic dimensions of the rogue wave, Kidd froze and stood looking up at it, and passed sentence on himself as "simply lost."

It was his close friend Doc Steigel who at the last moment grabbed Kidd and pulled him down. Kidd hit the deck hard and instinctively reached for a handhold. But he could find none.

Doc Steigel did find a hold, and, with his incredibly powerful grip, managed to hold both Kidd and himself fast to the deck as the black, suffocating wave drove in over them. The wave caught the *Saint Patrick* broadside, lifting and rolling her 200-ton bulk effortlessly as it exploded along her full length.

The blow from the thundering mountain of ocean water knocked the *Saint Patrick* on her side and nearly flipped her over completely. For a time, Kidd and Steigel felt the deck tilt almost vertically beneath them. The wave smashed the ship's stern handrails into grotesque tangles of useless steel tubing. So tall was the wave that seawater poured down her smokestacks (some twenty-five feet above the ship's waterline) and flooded into her engine room. Worse still, the wave's force smashed through a number of the ship's wheelhouse windows and poured inside, shorting out most of the radios and navigational equipment and knocking the skipper and several crewmen to the floor.

When the flooding salt water reached the auxiliary engines, all lighting throughout the *Saint Patrick* flickered, then failed, and her interior was doused, leaving her in a coffinlike darkness.

It was then that Doc Steigel panicked. "We've got to get off this boat, Bob!" he yelled to Kidd. "We've got to get off this boat! She's going to go down!"

"No, no, Doc! She's coming back up! See? She's responding! You can feel it rolling," Kidd yelled back.

Feeling their way through the darkness, Kidd and Doc climbed up the side of the wheelhouse to the life raft. Unable to go any farther, Kidd turned to Doc.

"Doc, I can't see the raft! We've got to get inside the wheelhouse! I know they bought two new flashlights in town."

Halfway down the side of the *Saint Patrick*'s superstructure, Kidd pulled open the side door of the wheelhouse. Listing as the ship was, it was like opening a cellar door. But there the similarity ended, for when they stepped inside the wheelhouse they were confronted by a nightmare of runaway emotions and physical wreckage. The scene terrified Doc. The ceiling was dripping with seawater. Navigational equipment torn from its mounts now dangled and swayed on the ends of uprooted strands of loose wiring. A door hung ajar. And there on the slick, slanting floor of the dimly lit wheelhouse lay the ship's skipper, William Green. He was ensnared in wiring that had been ripped from the ceiling by the wave, wiring that was now sending intermittent shocks of electricity through him. Green was struggling wildly to free himself.

At the same time, crewman Wallace Thomas was standing with his feet braced wide on the steep, wet floor of the wheelhouse, frantically sending out Mayday calls.

"Mayday! Mayday! Mayday!" he yelled repeatedly. "This is the F/V *Saint Patrick*! This is the *Saint Patrick*!" But there was no answer.

Then the skipper began yelling, "We gotta get off this thing! We gotta get off!"

Up until then, Doc Steigel had stuck close to Kidd, but when Doc heard his skipper yelling for everyone to abandon ship, he blanched. He was positive the *Saint Patrick* was about to roll over.

"We have to abandon ship!" Doc screamed to Kidd. "We haven't got any power! We're going to lose it!" And he bolted for the deck below.

Kidd raced through the darkness after him; he caught up with Doc at about amidships. Doc was already in the process of abandoning ship. He had one leg dangling over the side when Kidd grabbed him.

"No! No, Doc! The boat's going to stay up!" yelled Kidd above the howl of the wind.

"No! It's going down! It'll suck everybody down with it!"

"No, man! Stay with the boat!"

Suddenly, Doc shoved Kidd away. Kidd grabbed for Doc,

pushed back and this time when Doc tried to shove back, the two men began wrestling with one another. Tall and generally good-natured, Doc had been a logger in Washington State and was especially strong; in fact, he was one of the most muscular men Kidd had ever met. Kidd had even teased Doc, saying he had muscles in his eyelids. And now Kidd discovered that there was no way to stop him. The best he could hope for was to delay Doc's jump, at least long enough to try and reason with him.

But Doc Steigel wouldn't give in. And so Bob Kidd finally shouted, "All right! We'll go together! But first let me find a piece of rope so we can tie ourselves together. Calm down! Just calm down!"

But Doc refused to wait any longer. He leapt overboard before Kidd could stop him.

To Kidd, the world awaiting them overboard was a miserable black "hell-hole," a void of wind and darkness, of blowing spray, huge seas, and unending fear. At least on board ship, Kidd contemplated, there was some protection from the wind and elements. Yet, caught up in the panic of that moment, staying on board seemed equally terrifying.

Like Doc Steigel, Kidd now believed that the *Saint Patrick* was sinking. It was difficult to tell in the howling wind and spray and near-total darkness, but the ship was still listing badly. The thought of "getting sucked down with her" played heavily on Kidd's mind.

"Two people stood a lot better chance together" than one man drifting by himself, Bob Kidd reasoned. And so, grabbing a piece of rope line he found nearby, he, too, vaulted overboard.

Kidd struck the water feet first, going under briefly. Then he felt the buoyancy of his survival suit carrying him back to the surface. He squinted into the darkness, spied Doc Steigel, and swam to him, tossing him one end of the rope. Steigel and Kidd wrapped the rope around themselves and, afraid of being crushed against the bouncing hull of the *Saint Patrick,* they swam for their very lives.

Once clear of the crushing hull of the ship, they drifted. And as they did, they could see a light on top of the *Saint Patrick,* as if "someone was waving a spotlight." Moments later, though, the light seemed to go out. Driven by eighty-knot storm winds, the dark, listing form of the *Saint Patrick* rode quickly across the tall ocean waves and disappeared into the night. And then there was only an inky blackness and the roar of invisible breakers collapsing far off in the lonely night.

There was no turning back now, nor time for reflection. For only seconds later, a noise like that of a locomotive came barreling down on them from out of the darkness. The collapsing wave top sucked Kidd and Steigel up in its turbulent vortex, burying them under tons of seawater, and tossing them about. Regardless of their efforts to hold on to the line connecting them, the force of the wave tore the two clinging men loose.

Wave after wave battered Steigel and Kidd. With each new blow, they found themselves somersaulting end over end in the suffocating darkness inside the curl of the wave. And each time, they fought to free themselves from the current and make their way through the black depth of sea to the surface, where they could once again cough and hack and take in deep lungfuls of the precious air. Completely unable to see each other in the wintry darkness, the two crewmates caught their breath and immediately began yelling to each other. Then, listening for a reply, each swam in the direction of his crewmate's voice. Lost in the pitch-black night and muting roar of the bellowing Alaskan storm, it was the only way they had of locating each other.

"Doc! Doc!" Kidd yelled. "Where are you?"

"I'm over here! I'm over here, little buddy!" Doc answered, sentiment mingling with philisophical defeat in his tired voice.

When they found each other, Kidd said, "Doc, this rope is going to tangle us all up. We'll just lock our arms together." "And blow up your air bag [inflatable collar]!" Kidd shouted finally. Then, discarding the line, they interlocked elbows; lying on their backs, they tried to drift up and over the pummeling seas.

Doc began to calm down then, and though they had to yell to be heard, the two of them even started joking around.

"Gee, I wish I had a cigarette," hollered Kidd.

"I've got a cigarette if you've got a match," Doc replied. "We can at least have a smoke out here."

It was a way of belittling their predicament, a gesture of denial that helped keep them going.

Drifting along blindly through the night, they held on to each other and listened to the building storm. With increasing frequency, the towering and invisible rollers rumbled near and collapsed down over them. The wave wash would drive them under and tear them apart, forcing water into their suits. And each time Kidd and Doc found themselves separated by the rushing torrents, they would surface, screaming once again to help locate each other, lifting their youthful voices in the wild and inhospitable darkness.

Late that night, Doc Steigel began to develop a leak in the hand of his survival suit. It was a new suit, Kidd knew. But Doc couldn't seem to get his floating collar to work right. Now that his suit was beginning to fill with seawater, he was having greater and greater trouble keeping his head above water.

"He got close to me," recalled Kidd later. "I tried to feel for the cord that would inflate the collar, but I couldn't do a whole lot with our arms interlocked. It was tucked in or something, and in the darkness I couldn't find it."

"Bob," said Doc Steigel, "I've got a leak in my suit! I'm going to fill up with water!"

"Don't worry about it, Doc!" Kidd yelled back. "You won't sink! These suits can fill up completely with water and you still won't sink. So don't worry about it."

When yet another immense storm wave came upon them, Kidd and Steigel braced themselves as best they could. They turned their faces away from the icy onslaught, locked arms, and held on, but when the unfathomable tonnage of the huge wave broke over them, it again tore the two young men apart and carried them away on separate courses into the night.

When Kidd found his way back to the surface, he caught his breath and then began screaming for his friend. Seconds passed. Then he thought he detected the sound of something. Yes, it was indeed the voice of his friend Doc. But it came to him noticeably fainter now, from well off in the darkness. And this time Doc's voice carried with it a tone of hopelessness. As Kidd swam nearer, Doc cried, "I can't make it! I can't make it. I've got a leak or something. I'm not going to make it."

Another wave pounded over them, and when they surfaced again, Kidd quickly located Doc and immediately set out to encourage him.

"Don't worry about it, Doc," he yelled. "Just keep your arm out of the water!"

But it seemed to do little good, and the situation only grew worse. Doc was certain that a flood of the deadly cold seawater was pouring into his suit. Doc, Kidd knew, was giving up hope.

"I don't think I can make it," he hollered to Kidd a short time later.

For the first time, then, Doc and Kidd began to lose contact with one another. Bob Kidd soon found he "was having to scream and scream" just to get Doc to answer. And as an unending series of waves continued to batter them, it seemed to Kidd that Doc wanted less and less to get back together.

"You are stronger than I am!" Kidd encouraged. "I can't make it without your help, Doc!"

"You can make it, little boy!" came Doc's weak and fading reply.

"But Doc, I can't make it without you!" retorted Kidd.

Bob Kidd could barely hear Doc's last reply. "You can make it, little buddy," he said.

Kidd screamed back, blew his whistle insistently, then waited and strained to hear. The only sound that came to him was the waves. Doc Steigel was never seen or heard from again, nor was his body ever found.

When his deckmate and friend finally drifted away, a dreadful loneliness struck Kidd. Utterly alone, he drifted through the tar-black night, all the while waiting and listening, ever vigilant, for his friend's voice.

Kidd could feel his own hope fade and his willpower begin to falter. From the moment he had abandoned ship, he had felt he was doomed. He thought about death and about how long he might survive under such impossible conditions. He was about 100 percent sure he would never make it out alive.

For an incessant series of waves continued to roll upon him from the impenetrable darkness, striking without warning and in close formation. There was little time to think. It was like being caught aboard a watery roller coaster careening out of control. Each wave tossed him through what seemed to be hundreds of loops. And in spite of his best efforts, icy seawater now began to force its way into Kidd's survival suit.

With Doc gone, Kidd felt doomed. He began to weep and

plead aloud as the waves lifted him into the night. "I don't want to die! I don't want to die!" he said aloud, over and over.

Then he remembered how Doc had given in to the same despair. Don't think like that! Snap out of it! he scolded himself. If you panic, you're going to end up going the same way.

I'm breathing. I'm alive, he thought, calming himself. I'm going to stay with it as long as I have to.

Then he prayed. He thought of his mom, his family, and of the events in his life. If I'm dead, I'm gone, he reasoned. But I'm not the only one who'll get hurt. My family will, too.

Doc was gone. Kidd hadn't realized how much he would miss his company. But there was little time to mourn.

"Every so often you could hear a wave starting up and just coming," he recalls. "You couldn't see it, but you could hear it. It sounded really, really loud! And whenever I heard one, I'd put my hands over my face and get ready for it. It would really roll me around; then I would come out of that, having taken in a lot of water. It was hard to get my breath back after swallowing so much seawater. And it seemed like before I could really start thinking again, another one would come along."

Despite the fact that he was wearing wool long johns, a pair of cotton socks, and a red flannel shirt underneath his survival suit, Kidd soon grew "cold as a bastard."

During the night, he began to shake uncontrollably. So Kidd withdrew his legs and arms and huddled up in the fetal position. "I was just in the belly of the suit," he recalls. "I don't know how I did it."

Kidd was clinging to the hope that his crewmates had remained on board the *Saint Patrick* and that they would rescue him in the morning. They'll be out looking for me then, he told himself.

Dawn broke to a world of "big, high rollers." A thick fog bank had slid in over the area, and Kidd worried that yet another storm was building. There was no sign of the *Saint Patrick*.

Still adrift in the icy Gulf of Alaska water, Kidd began thinking "these stupid things." He worried that he was growing delirious. The visibility was poor, and he was "pretty well out of it" when he rose over the top of a large wave and glimpsed through a break in the fog, what he believed to be the tip of an island in the far distance.

The land was too far away to make out any trees—or any detail at all with any certainty. He figured land was at least ten miles away. And in that same moment, he put his arms and legs back inside the survival suit and started swimming.

"Oh, thank you, God. Thank you, God," he said aloud. Lying on his back, he backstroked toward the sight, pausing only long enough to relocate land periodically and to correct his course.

If I can see it, I can get to it, he told himself. Nothing's going to stop me now! And so for the next nine hours, he swam. The seas tossed him one way and then another, and he "fought the current the whole way."

Feeling exhausted and a little delirious, Kidd paused to catch his breath; he found himself drifting through a series of large wave troughs. Shortly, a large seagull arrived and began circling him in the windy sky overhead. Floating through the air only a few feet away, the bird seemed to be studying him. He felt moved to defend himself aloud to the bird. "Hey, don't look at me," he shouted above the wind. "I don't know what I'm doing out here. I don't want to be doing this, you know." The seagull seemed satisfied with his explanation, and soon flew on.

Late that afternoon, Kidd approached Afognak Island. Several hundred yards offshore, he noticed "these green bulbous things" leaping from the surf and then disappearing once again into the sea. He realized they were clumps of kelp. Kidd studied the shoreline carefully and quickly realized that there wasn't any safe place in the entire area. Caught in the current, he was being dragged directly toward the terrifying formations ahead. Fourteen-hundred-foot cliffs rose up before him. The black, wave-slickened wall of granite looked totally impassable. As he drifted in closer he watched the towering explosions of the driving surf pounding hard against them, heard the deafening roar of impact, and saw the furious white tonnage of draining water fall into the sea and get sucked out again.

The deadly vision terrified Kidd. In a moment of panic, he attempted to swim away. But the current proved too strong. The next thing Kidd knew, he got sucked inside the water. "I don't know how far out I was, but it took me all the way to shore. It was tearing at me. I was underwater. The next thing I knew, I was hitting against the rocks and getting pulled back out. . . . I was constantly underwater . . . and being beaten against the cliffs. And I would try to grab a handhold on the bottom and hold on with one hand at a time."

Sucked beneath the surface of the sea, his suit quickly filled with water. Kidd felt like a man caught inside a balloon filled with water. The power of the thrashing surf was unimaginable.

Then, as Kidd felt himself slammed once again against the cliffs, he realized that if he was to survive, he had to get rid of the suit. "It was drowning me!" he recalls.

Submerged in the water, clinging to any handhold he could secure on the bottom, Kidd unzipped his survival suit far enough to free his arms, and the powerful shoreline currents literally stripped it from his body. "It was gone in a second, just gone," he recalls. "I could feel the weight of it leave me, because it had filled up completely with water."

Kidd surfaced then, and when the next wave catapulted him up against the cliff face, he grabbed new handholds in the rock and held on. Each time a wave roared ashore, it would explode up the sheer rock face of the cliff and over Kidd. But he would freeze fast. Each time the water threatened to strip him backward and drag him off into the surf, Kidd would clutch the rock face of the cliff with a death grip and fight to remain in place against the drenching force of each new breaker.

During the pause between barrages, Kidd began to climb. Terrified and exhausted, he picked his way toward higher ground as quickly as his numb body could move. He climbed slowly. Twice he was nearly knocked from the face of the cliff, but somehow he managed to hang on.

He felt encouraged when he rose to an elevation at which only spray and not the wave itself was washing over him. But not until Kidd had reached a height of at least fifty feet above the surf itself did he finally break free of the deadly wash.

Continuing his climb, he clambered atop a pinnacle of rock that offered a panoramic view of the area. Looking down along the shoreline, he spotted the small stretch of beach he'd originally aimed for. He'd missed it by no more than fifty yards. He contemplated climbing back down. But a single slip would mean almost certain death. Facing such a cold and deadly surf in his weak and hypothermic condition without a survival suit seemed impossible.

Looking skyward then, Kidd found himself faced with hundreds of feet of sheer cliffs. Jagged pinnacles of windswept rock soared above him, while below him the deadly surf pounded in against the bottom of the cliffs. Up or down, the climb before him appeared physically impossible.

He knew he had only one way to travel, and that way was up. Standing fully exposed to the 20°F. chill of the twenty-knot wind, Kidd "knew he had to either go up or else stay there and die of exposure."

Without shoes, and clad now in only his long johns, cotton socks, and flannel shirt, Kidd sat down to catch his breath, rest his shivering body, and take stock of himself. He was soaking wet, nearly naked and severely hypothermic. He knew that if he was going to survive he would have to somehow scale the cliffs, rising high overhead, and get into the woods and out of the biting cold of the wind.

And so he began to climb. Each time he chose a route, he would climb full of hope, but he would "get only so far" before the handholds gave out. "The next handhold would be too far to reach," and he would climb wearily back down and pick another route, then slowly venture upward again.

By the end of the second unsuccessful climb, his socks were reduced to mere pieces of cotton stuck to the sides of his feet—feet that were aching and badly swollen.

He would make one final attempt. Slowly, carefully, Kidd picked his way up the face of the cliff. Several hundred feet above the pounding roar and crash of the surf far below, Kidd crawled out on a dizzying ledge. And there he came face-to-face with a gaping crevasse stretching a full ten feet wide directly in his path. There was no way around it. He felt like giving up. It was ridiculous, unfair. He couldn't take it any longer.

"Why?" he asked the Lord aloud. "Why are you doing this to me? Why would you bring me through all this, just to let me die here? Why place another obstacle in front of me? I can't take it anymore!"

Dripping wet and completely exposed to the knifing cold of the wind, Kidd saw only rock and more rock rising above him as far as he could see. He was too weak, he knew, to climb back down again. But to attempt a crossing might well mean a fall and certain death on the rocks below. Kidd knew he was too cold to continue much longer, and too fatigued to attempt a climb back down.

Freezing cold, "totally exhausted," and realizing that he could go no higher, Kidd sat down on the narrow ledge. His feet were swollen and severely frostbitten; their sides were split and bleeding. As he sat there, he tried to warm them by covering them with moss torn from the face of the ledge-rock.

At that moment, a brightly painted U.S. Coast Guard helicopter flew past, clacking loudly as it went by at an altitude almost exactly even with Bob Kidd's cliffside perch. Kidd almost panicked with excitement. It seemed he could almost reach out and touch the chopper. He could see the faces of the two helmeted men seated in their pilots' seats; they appeared to be engaged in conversation. Tearing off what remained of his red flannel shirt, he waved it frantically back and forth overhead and screamed.

"The helicopter flew exactly level with me, and I could even see the guys in it. I started waving at them. And, of course, I no longer had the suit on, and they didn't see me. The helicopter was very close. . . . I was looking the pilot right in the face!" But the helicopter passed on, never to return.

Shaking from the relentless chill of the wind and racked by a hopelessness that went beyond words, Bob Kidd slumped down and "cried like a baby."

When Kidd looked up again, he took in a long, flat slab of

rock about fifteen to twenty feet in length. The slab was on the far side of the crevasse, slightly below him, about ten feet away. It was imbedded in the far cliff face at about a fifty-degree angle. He couldn't see any hand grips on the rock's rain-slicked surface, and he knew that if he were to slide off, the unwelcome ride would carry him into a deadly oblivion, a free fall straight down onto the craggy rocks hundreds of feet below.

For a while, he sat and studied the rock slab and the terrain above it. He came to the conclusion that should he somehow survive his leap across the crevasse, the rest of the journey to the protective forest above looked relatively simple.

To remain where he was, fully exposed to the deadly Alaskan wind, with darkness closing, meant that he would soon die. Kidd was certain of that. He stood up then. He could hear the thunderous reports of the surf exploding far below. It was enough to drive a grown man to the edge of sanity.

Without further contemplation, Kidd turned, and dived headfirst across the gaping crevasse. He stretched his cold-ravaged body toward the sloping face of the rock slab with everything he had, hit hard, and nearly fell. The impact knocked the wind from his body, but somehow he "just stuck to the surface," and his feet and hands grappled for notches to cling to.

Bob Kidd lay there clinging to the edge of the wet and sloping face of the rock slab for several minutes before he could summon the strength to pull himself up and begin climbing again.

Another hour of such efforts propelled him over the top of the fourteen-hundred-foot cliffs. There he found some freshwater. In his cold and weakened condition he sensed that consuming a large rush of cold fluid could lower his already badly chilled body temperature to a deadly one, and he began to sip at it ever so slowly. Then he sat and attempted to gather his faculties. He had no knowledge of the immediate area, no idea how large the island was; he couldn't remember if there were any Eskimo villages on it or not.

The country on top of the cliff was wild, untouched wilderness, and large portions of it were covered by stands of spruce trees. Some of the largest brown bears in Alaska proliferated on Afognak Island; he would have to keep a constant watch out for them. But he'd also heard that there were hunters' cabins scattered here and there around the island. And with that in mind, he began "walking and stumbling," in search of help.

As he traveled, he glanced down and took stock of his feet. Discolored from frostbite, they looked as long and wide as toy sailboats. And though he stepped on sharp objects, he could no longer feel them—not sticks, nor rocks, nor the ice on the frozen ground.

Later as he traveled, he noticed a large opening in the flesh of his foot. It didn't hurt, but, even more strange no blood was coming out. Then a realization struck Kidd. "My feet were frozen solid. . . . There was an opening and no blood was coming out. I could look right in there."

With darkness closing in and visibility fast diminishing, Kidd decided to bury himself in whatever he could find. His body shook incessantly, and he was colder than he could ever have imagined, but he did his best to save what little body heat he had left.

Then he came upon a windfall, a spruce tree that had been toppled by the wind or old age. The roots had pulled partially free of the ground, forming a small natural cave underneath. Kidd wasted little time in gathering pieces of moss and laying

them over the screen of broken roots, and in so doing succeeded in making himself a "little hut, a shelter out of the
wind."

Exhausted and feeling faint, he crawled inside and began
pulling moss and leaves, limbs and grass over himself. Soon
he felt a heavy sleep coming on. Strangely, the incessant cold
that had for so long mercilessly racked his body now somehow
seemed unimportant. Instinctively, he knew that he had "cre-
ated a little grave" for himself. He would crawl in, drift off,
and just let life go.

Huddled down in his crude makeshift shelter, Kidd suddenly felt the complete absence of cold. He took it as an omen
of his approaching end, and so he set about to make peace
with God and himself.

The next thing Kidd knew, it was daylight, and a helicopter
was whirling past overhead. The Coast Guard was looking for
him, he knew. But he also knew that if they didn't find him
that day, it would be over for him. Kidd rose stiffly from the
crude shelter he had made and soon spotted a stretch of sandy
beach down through a stand of spruce trees, and he made a
straight path toward it.

He knew he'd have to get down onto the beach. If there
were other Coast Guard helicopters, that's where they'd have
the best chance of spotting him. As he worked to close the
distance, he felt the desperation only a dying man can know.
But with his reserves of strength nearly gone and his frozen
feet refusing to respond underneath him he was forever collapsing and rising as he stumbled forward toward the beach.

Once on the beach near the water, Kidd took in what looked

to be an entire armada of rescue craft. There were fishing boats and U.S. Coast Guard SAR ships scattered across the horizon. The skies seemed to be filled with planes and helicopters.

Kidd recalls, "I would get up and wave, but I would fall back down because I couldn't stand up by myself. I would get up and go back down over and over again. I did that for quite a while; then I went behind the bush and stayed there, and I was crying because I had tried everything and the planes went right over me and didn't see me."

He would never have believed he could have grown colder than he'd already been, but as he stood entirely exposed to the piercing cold of the November winds, Kidd did just that. He knew that the wind and cold would soon kill him, and so he kept stumbling, hurrying back and forth between the beach and the sheltering windbreak of the trees.

"Then I went back on the beach and a helicopter came in, and I knew they'd seen me. They had to come about fifty yards down the beach." Terrified that something else would go wrong, Kidd staggered toward the hovering chopper on his own wobbly legs.

"I saw them lowering the basket . . . and I smashed into trees, I fell over trees, I fell over rocks, and then I fell into the basket, yelling, "Take her up! Take her up!"

Kidd was in critical condition when the crew of the helicopter lifted him aboard. His body temperature registered a mere 81° as Coast Guard attendants began their battle to save Bob Kidd's life. Flown immediately to the Coast Guard cutter *Confidence,* which was waiting nearby, Bob Kidd was treated by Dr.

Martin J. Nimeroff, a Coast Guard physician and hypothermia expert. Having lost the majority of his toes and the flesh on both feet to frostbite, it would take Bob Kidd nearly a decade to recover fully from his ordeal.

Over the next few days as he lay recuperating in the intensive care unit of the Kodiak village hospital, Kidd received word that shortly after he and Doc Steigel had abandoned the *Saint Patrick,* the ten remaining crewmen on the ship had also jumped overboard. They, too, had spied land at first light and had set out to reach it. Some died that first night. Others died en route. Most of those who managed to reach Afognak Island perished on the rocks going in through the surf. Only one crewman, a young man named Wallace Thomas, had survived; Captain William Green and the majority of his crew had become the victims of inexperience, bad luck, panic, and tragic incompetence. Ten of the twelve crewman on board had been lost.

Then Bob Kidd learned the most incredible news of all. The 155-foot *Saint Patrick* never did sink. Drifting unmanned through the howling storm and black of night, she had remained afloat. A group of fishermen had discovered her floating the next day. The ghost ship *Saint Patrick,* Bob Kidd was informed, was being towed back to port at that very moment.

Several years later, in an Anchorage courtroom, lawyers representing the relatives of the crew of the *Saint Patrick* won a $13.5 million lawsuit for negligence against the owners of the *Saint Patrick.*

This tragic incident would play a vital role in the 1986 congressional hearings in Washington, D.C. Arranged through the efforts of the grieving parents of yet another sailor lost unnecessarily in Alaskan waters, the hearings helped focus national attention on the fight to upgrade safety standards aboard commercial fishing vessels in U.S. waters. As a direct result of those hearings, U.S. Maritime Law was heavily amended in 1988. Survival suits for each crewman, life rafts, radios, and extensive crew training are now required aboard all commercial fishing vessels underway in U.S. waters.

IN THE PATH OF A
MIRACLE

It was the fall of 1979 and the high-stakes Kodiak king crab season was already well under way when skipper Gary Marlar, owner of the fifty-foot fishing vessel *Harder,* fell ill and had to be hospitalized. The fall king crab season was the single-largest moneymaking season of the entire year for fishermen in the Kodiak area. Marlar knew it only too well: That single season could make the difference between success and failure—between fiscal solvency for that year and years to come, or outright bankruptcy. But now he found himself lying in a hospital bed, attached to a lung machine, effectively eliminated from the season at hand.

Several weeks before, Marlar and his crew of two had baited and set their king crab pots on some little-known chunk of the local fishing grounds just a few miles off Kodiak Island. When Marlar returned to shore, however, he contracted pneumonia and had been forced to enter the hospital.

Gary Marlar had been there for the better part of two weeks when his doctor, Kodiak's James Halter, approached him in his hospital room. "Well," the kind doctor said to Marlar, "I'm going to let you out of the hospital—on one condition and one condition only. You must promise to go right home and go directly to bed and get some more rest. You are still quite ill."

"But the king crab season's already begun," Marlar complained. Helpless to make his case, he finally agreed. Still, "the crab pots had been out there lying on the bottom baited and fishing for two full weeks," recalls Marlar, "and I wanted to get out there and get with it." He had hardly finished signing his hospital release papers when he called his crew and ordered them to meet him down on the Pan Alaska Cannery docks, where the *Harder* was tied. They would shove off for the offshore king crab grounds that very night—at 4:00 A.M. sharp!

That night, one of his two crew members failed to show up for the trip; Marlar could have called it off, but he decided to go anyway. Moreover, in the middle of the night, another young man appeared, hungry for the chance to work aboard an Alaskan crab boat. He had no experience, and he didn't own a survival suit. But he said he would work for free if he had to, just to get the experience. "If I don't work out, you don't even have to pay me!" the ambitious young man told Marlar.

As a longtime skipper, Gary Marlar considered himself a reasonably good judge of deckhand material. As for this young comer, he actually liked what he saw. "He had a manner about him that indicated he wasn't just some dock bum," Marlar recalls. But for some reason, Marlar found himself refusing the offer. "No," he said finally.

When the young man became upset, Marlar encouraged

him to look him up again when they got back to the dock there in Kodiak. Neither Gary Marlar nor the disappointed man left on the dock would understand until much later how providential Marlar's decision had been.

Jeff Hinshaw would comprise Gary Marlar's entire crew for this trip out to sea. Hinshaw was a college student from Bend, Oregon; Marlar had met him on the waterfront during salmon season the summer before, when he delivered a load of salmon to the B & B Cannery on Kodiak's Cannery Row. He'd noticed at once that the young man was "one hell of a fine worker." When the salmon season ended, Marlar looked him up and hired him on as a crewman.

It was overcast and rainy, blowing thirty knots or so, when Marlar fired up the engine aboard the fishing vessel *Harder.* The first gray inkling of dawn was already showing as they idled along the waterfront and out of the port of Kodiak.

Once in the fishing grounds, Gary Marlar was forced to leave the wheelhouse and assist Hinshaw out on the back deck, not an uncommon practice on smaller, seiner-sized fishing boats. There was only a three- or four-foot chop at the time and, as they worked, the *Harder* lay side-to in the close-cropped seas, rocking back and forth with a casual indifference.

Marlar ran the hydraulic controls as Jeff Hinshaw stood and, hand over hand, coiled the flashing yellow loops of 5/8-inch polyester line into neat circular piles. The whining steel block winched crab pot after crab pot from the ocean bottom some seven hundred feet below.

Both men were working on deck when, around two o'clock that next afternoon, "all hell broke loose."

Before, the *Harder* had appeared to bob effortlessly through choppy seas and bowl-shaped wave troughs. But now as they worked, the ship suddenly took a roll onto her starboard side. Marlar watched the handrail slump down into the sea and remain there. Seawater began to spill onto the deck.

Aw, hell, thought skipper Gary Marlar, we must have one awfully heavy pot here! We're on 'em! This one must be plum loaded with crab! But when the pot finally broke the surface, he was surprised, for there were only a couple of king crab inside. Something was wrong. In an effort to right the *Harder,* Marlar released the pot to the depths and tossed the crab pot line and buoys back over the side, but, to his dismay, the *Harder* refused to budge.

Jeff Hinshaw headed down into the engine room, but before he could stop his momentum, he had waded out into a waist-deep flood of seawater. They were in big trouble, and he knew it. "I've got water up to my waist down here!" he yelled.

"Go grab your survival suit and get out on deck!" Marlar yelled back. Moments later, Hinshaw appeared, carrying his suit, and climbed up onto the rolling surface of the ship's bow deck.

For the engine room to flood with seawater so quickly, something had to have gone desperately wrong, Gary Marlar knew. So much water had poured in so quickly; perhaps a plank on her wooden hull had given way, or a seam on her aluminum live tank had split. But with the *Harder* listing so sharply, there would be no time to locate the problem. Marlar raced up into the wheelhouse, grabbed one of the prongs on the forty-inch wooden steering wheel, and spun it hard to starboard. He would attempt to save the listing vessel by using the force of the oncoming seas to right her.

But the *Harder* refused to budge; she remained slumped over onto her starboard side. Gary Marlar could hardly believe the sudden and nightmarish turn of events. Yet he knew the situation was critical. He grabbed his survival suit, stuck his arm through the handle loop, and slid it up to his shoulder. Then he reached overhead to grab the radio microphone. He would make a Mayday call to the U.S. Coast Guard base on nearby Kodiak Island before another moment was lost.

There was only one wheelhouse door leading to and from the outside, and that door was now submerged. With seawater rushing into the galley and staterooms behind him, and more water blocking the doorway, Marlar was trapped. He struggled to hold his footing on the near-vertical slope of the wheelhouse floor. He knew it was imperative that he get a Mayday call for help sent off before the ship went down, but there was no time, and the physical act of standing soon became impossible. For at that very moment, the *Harder* rolled completely upside down.

Faced with the prospect of becoming entombed there and being carried straight to the bottom, Gary Marlar turned and dived down into the dark green rectangle of ocean water filling the doorway.

In an effort to remain clear of the inverted wheelhouse of the sinking craft, Marlar stroked straight for the bottom. He was too terrified to feel the initial shock of the wintry Gulf of Alaska water. But he had gone no more than ten feet when he felt the hood of his rain jacket snag on something. He struggled wildly against the unexpected entanglement, but it held him fast. He could no longer go up or down. Feeling his supply of oxygen waning, he battled mightily against the unseen impediment that now seemed determined to drown him.

Submerged in the brackish depths, he could feel his chest tighten and his lungs begin to burn. The lead weight of his boots was pulling him down. Soon one thought was screaming through his mind: Oxygen!

Then, as he struggled, he felt the hood of his jacket break free. Determined to avoid any further such terrifying entanglements, Marlar forced himself to swim still farther down into the depths. Finally, convinced that he was clear of the lunging, inverted form of the wooden wheelhouse, he stopped his descent. He pivoted in the water, spied the dull gray light overhead, and struck out for the surface.

Gary Marlar broke through into a world of fresh marine air and inhaled reflexively. Clad in nothing but his work clothes and rain gear, he dog-paddled there on the surface, trying to gather himself. His first conscious thought was for Jeff Hinshaw. Marlar could not see him; he was "worried sick" that the young man had not escaped from the *Harder* alive.

He could see the wet, rolling hull of his overturned crab boat few yards away. But there was no sign of the youngster, and as he searched, Marlar felt the tidal currents pulling at him. He fought against the uncomfortable slap of the brutally cold waves breaking over him.

He swam to the nearest edge of the overturned vessel and held on. Seriously weakened by his two-week hospital stay and bout with pneumonia, Marlar knew he had to rid himself of his flooded boots and rain gear and get into the protective folds of his survival suit as soon as possible.

When Marlar reached the hull, he called aloud for his crewman. He received a reply almost immediately.

"I'm over here!" called Hinshaw from somewhere on the other side of the hull. "Where are you?"

Marlar felt his heart jump with relief. "Over here by the stern!" he shouted. "Swim to the stern if you can!"

When Marlar spotted the young man, he was treading water on the opposite side of the boat. "Jeff! You all right?"

"My boots!" yelled Hinshaw above the wind and waves sloshing against the hull of the *Harder*. "I'm having a hell of a time swimming! I mean, it feels like someone is grabbing my ankles and trying to pull me down!"

"Hey, let's climb up on the keel," Marlar suggested. "We can get our boots off, anyway. They'll weight us down. I'll help you with your rain gear, and you can help me with mine. Then we'll try to pull on our survival suits."

Through it all, the ship's diesel engine had somehow kept running, and it continued to now in spite of the fact that the vessel was completely inverted. Her propeller was still spinning in air, so Marlar and Hinshaw took great care as they swam near to avoid the lethal flash of the spinning propeller blades.

Now the two fishermen attempted to scale the wave-washed hull. They approached from opposite sides and met each other on the keel at about midships. They rushed to remove the deadly and cumbersome bulk of their boots. Their rain gear would come off next, to be quickly replaced by the all-critical protection of their survival suits.

Jeff Hinshaw asked Marlar one weighted question then. "Did you get a call out to the Coast Guard?"

Marlar shook his head. "No," he said. "There wasn't time."

Marlar watched Jeff Hinshaw's demeanor change. The young man's face winced at the news. A dark realization now reflected in the young man's eyes.

They had hardly succeeded in removing their boots when the *Harder* began to sink out from under their feet. The ship went down in seconds, stern-first.

Marlar yelled to Hinshaw. "We'd better swim off this and

get clear of the boat! We don't want to get sucked down with her!"

Hinshaw leapt off one side of the boat and Marlar dived off the other. Both stroked hard for the distant horizon. They were about forty feet apart when the *Harder* went down for the last time. When she had finally disappeared, gray ocean waves quickly swept over the site. It was almost as if she had never existed.

Though he had seen it with his own eyes, Gary Marlar could hardly believe it. For the owner and operator of the sinking vessel, it had been a "love-hate relationship from the very beginning." The *Harder* was, first of all, a "pain in the ass." He had poured everything he had into purchasing her, getting her into working shape, and keeping her that way—all the time, money, and passion he could summon.

And as he drifted there on the water on that lonely gray October day, he felt a deep sadness. Part of him was "gone forever." He felt as though all of his efforts and hopes and dreams had gone down with her. After all the years of painstaking repair and care, the *Harder* had taken only about three minutes to sink.

Hinshaw and Marlar swam toward each other and worked feverishly to help remove each other's jackets and pants while treading the icy seawater. Time was of the essence

They removed the soft foam fabric of their survival suits from their bags and struggled to climb into them. But they "kept drifting apart," and then their ability to aid in each other's individual struggles was interrupted again and again as they fought to close the growing gap between them and get back together again.

Jeff Hinshaw managed to get into his survival suit and pull his zipper snugly in place under his chin. Then, using the clever purge valves set inside both legs of the suit, he was able to drain the majority of the icy seawater from his suit.

Gary Marlar, however, was not so lucky. Despite both their efforts, they were unable to close off completely the deadly, chilling flood of water inside Marlar's suit. Marlar soon found himself locked in a miserable and exhausting battle. The suit filled quickly with the insufferably cold water. The arms, legs, and torso of the suit bloated up like fat sausages, and his body shivered violently within.

Every time Marlar stopped treading water long enough to attempt to zip up the gap in the chin area of his suit, his head would go under. Without help, he quickly came to realize, he would "sink straight to the bottom."

"Hey!" Jeff Hinshaw hollered suddenly, focusing on something in the distance. "There's a can or something floating in the water there!" Hinshaw stroked quickly to the unidentified object. It turned out to be a five-gallon can of hydraulic oil.

The can, which was half-empty, would offer a certain amount of buoyancy. Hinshaw immediately began pushing it through the breakers toward the pale-faced figure of his skipper. Gary Marlar wrapped his numbed arms around it and held on.

Now, as they drifted, Marlar's attempt to call for help came up again. "Yah," Marlar said to Hinshaw, "I just wasn't able to make that Mayday call. I tried to, but the boat rolled right at that same time. I guess I could have stayed inside her and made the call. But I would have gone down with the boat, so I chose to dive out."

"Oh," began Hinshaw, "that's okay. It's only about three miles to the beach over there. It doesn't look too far for us to swim."

"Yah, I think you're right," agreed Marlar. "We can probably make it."

Jeff Hinshaw helped Marlar slide farther up onto the can. Clinging fast, with his stomach and chest sprawled across the length of it, Marlar began kicking with his legs. Hinshaw reached out and grabbed the can with one arm and, tugging as he went, stroked with his other hand toward the distant shore of Long Island.

Although his suit was still bloated, Gary Marlar felt a renewed sense of hope. For four hours, they paddled ahead determinedly, and if they grew tired, they never spoke of it.

"Yah, we're getting there. We're getting there!" Marlar said repeatedly as he swam.

Shortly before darkness descended, they spotted a U.S. Coast Guard C-130 SAR plane on the horizon. Those inside the plane were searching for a missing man in a skiff that had reportedly overturned. But looking down from so far overhead, they must have thought that the bright orange hoods of Marlar's and Hinshaw's survival suits were crab buoys.

Marlar and Hinshaw rocked forward in the water and raised their arms skyward. They waved back and forth, yelling, "Hey! We're right here! Hey! Heeeeeyyy!"

The plane passed by almost directly overhead; Marlar and Hinshaw watched helplessly as the plane flew on, disappearing on the far horizon.

Nightfall found them bucking into a nasty chop that was blowing straight offshore, directly into their faces.

Now, with the weight and chill of the seawater filling his suit, Marlar was growing dangerously cold. Inside his leaky survival suit, he was "just freezing" and growing weak from the impact of the piercingly cold water and his lingering battle with pneumonia.

Jeff Hinshaw finally stated the obvious. "I don't think we're getting anywhere."

"Yah," added Marlar, "I can't even see land over there anymore." They decided to stop and just hold on for a while.

As they drifted along through dimensionless darkness, Marlar and Hinshaw could make out the red and green running lights of other crab boats returning home from the offshore fishing grounds. Some were returning from fishing grounds hundreds of miles away. The two men could see them winding their solitary way through the buoy markers as they crossed the vast black void of night and the formless expanse of Chiniak Bay.

Though they were miles outside of the lane designated for returning ship traffic, the two drifting men hoped that perhaps, *just perhaps,* one of the returning vessels might possibly wander in their direction and rescue them from the miles-deep darkness and the slow and painful death by hypothermia that was drawing near.

After many hours of paddling and drifting through the cold and wintry darkness, Gary Marlar was played out. Never in more than twenty years of fishing and adventuring in Alaska had he experienced anything like the pain and numbness now disabling his body.

He could feel the swift tidal currents sucking him and his partner ever farther offshore—a fact confirmed by the vision of the red and green channel-marker lights and the buckshot pattern of twinkling shore lights on Cape Chiniak pulsing blue and white as they drifted past in the arctic darkness.

It was about midnight when Marlar first decided to drown himself.

Mental confusion is a common result of severe hypothermia. With the deadly, heat-stealing flow of icy Alaskan water in and out of his suit, Gary Marlar's body temperature was plummeting. Now his determination to get through the crisis alive, "no matter what," gave way to a feeling of calm, and then to one of gloom.

With his suit weighted and bloated as it was, "there was just no way" Marlar reasoned, that he was going to make it. And with the tidal currents carrying them farther and farther offshore, there seemed to be no use in trying.

Never before had Marlar seriously considered suicide. He'd never been given to self-destructive mood swings. But the pain was excruciating and endless.

He was trapped in a situation that was beyond his control. More than anything, he wanted it to stop. The cold was robbing his body of all heat, depleting him of what little strength he had left. It was a process that, if unchecked, would continue until the ebb and flow of blood and breath were stilled forever by rigor mortis. In his confused state suicide seemed a logical and even preferable choice.

Marlar decided to quit fighting. He would let go of his tenuous grip on the five-gallon oil can, stick his head under the surface, and remain there until he passed out. He could faintly recall reading an article about several deckhands in the Aleutians who described their near-death experiences while drowning. Or was it from hypothermia? The exact specifics were muddled, even lost to him now. The survivors had said that one or the other of those was the least painful way a man could choose to die. You just take in water, Marlar thought, and after a certain point, nothing hurts anymore and you just go, like in a dream, and that's it.

"Look, Gary, a boat's coming!" yelled Jeff Hinshaw, pointing tiredly.

Marlar looked in the direction of Cape Chiniak. Perhaps

ten miles off in the thick darkness, he could make out the lights of a crab boat traveling steadily along. But, like all the other ships they had spied traveling through the area, it was moving on a course well away from them as it crossed Chiniak Bay on its way into the snug protection of Kodiak's Saint Paul Harbor.

Still, Gary Marlar wanted to believe—was desperate to believe—that what his tireless crewmate was telling him was true. He wanted a quick end to the fear, the pain, and the hellish fatigue. And for the moment, the false promise of imminent rescue lifted his sagging spirits.

Without warning, the vessel turned and moved off into the night. The disappointment was "just too much" for Marlar. He'd "never felt anything like it." His heart seemed to free-fall right through his stomach.

As they drifted farther and farther from shore, the seas began to stand up and break over the heads of the two struggling survivors. The worst part was realizing that no one even knew they were missing.

"I just can't," Marlar said to Jeff Hinshaw in a downtrodden tone. "I just can't go on any longer. I'm going to take in some water and end it."

"You know, Gary," said Jeff Hinshaw finally, "if we pray together and ask God for help, sometimes it works."

"Hell, I'm willing to try anything!" shot back Marlar hopelessly.

Both men lifted their heads to the heavens then and began shouting their prayers aloud, asking God to spare them. While Jeff Hinshaw said a prayer in accordance with his Mormon faith, Gary Marlar looked up and yelled simply, "God, help us! Please help us!" Shortly his foggy mind came up with a brilliant conclusion. "We need a boat!"

Ever since the *Harder* had sunk and he and his crewman had drifted away, Marlar had fought back the guilt over having "not gotten off" a Mayday call before the boat went down. No one even knew that the *Harder* had gone down; no one had any idea of where they had gone, or where they were. Marlar was filled with a "lonely, useless kind of feeling." But now that they were praying, Marlar felt better.

They prayed aloud for "probably an hour or more." Then they fell silent, drifted up and over the waves, and were carried off into the darkness.

But the excruciating pain of hypothermia wore at Marlar, and before long a pall of gloom and hopelessness had returned. Marlar didn't mind dying, but he could not accept the thought of never seeing his wife, Debbie, and the kids again. He couldn't stand it. Looking up again, he pleaded once more with the heavens. "God, if you get me out of here alive, I swear I'll never come back out on the water again! I'll pick apples! Anything! Just help us!"

Jeff Hinshaw, too, began praying aloud again, asking for God's deliverance.

Now the drifting survivors felt the wind freshen, and the size of the ocean waves rose accordingly. "Every minute or so" as they drifted through the tall and building seas, an exceptionally large wave would break over their heads and a new, jolting measure of icy seawater would pour in through the zipper opening of Marlar's suit. Believing that the situation was hopeless, Gary Marlar dived down and tried to drown himself. Jeff Hinshaw swam over, jerked him back to the surface, and began railing at him. Marlar could hear the moral indignation and invective bursting forth in every syllable. "You've got no *right* to just quit and leave me here all alone!" he shouted, letting loose a torrent that mixed curses with words of exhortation and hope. "You can't give up! Just hang on! There'll be something!"

The sudden yelling and unexpected profanity coming from the religious young man helped lift Marlar out of his hypothermic stupor, if only momentarily.

Then Hinshaw's shouting took a different tack. "Hey, Gary, here comes another boat!" he cried.

"Bullshit!" shot back Marlar, the tone of his voice dripping with cynicism and anger.

"No, really! Take a look!"

"Jeff, just leave me the hell alone and let me die, will you?" Marlar pleaded.

"Hey!" shouted Jeff. "I see a mast light! There's a boat, and it really is coming our way!"

"Damn it, Jeff," Marlar said disgustedly, "don't do that to me anymore. I don't care if a boat's coming or not. I just can't take the disappointment anymore."

But when Gary Marlar finally did look up, it was to the growing vision of the lights of a fishing boat. He could hardly believe it; she was close enough that he could already make out the form of her bow. His heart leapt. The vessel seemed to be moving on a course that would carry it close by. Gary Marlar knew this was his last chance. He released his failing grip on the oil can and began swimming with everything he had left toward the approaching vessel.

Marlar judged the course of the oncoming vessel and swam directly into her path. And as he did, his tired heart was filled with two great and conflicting emotions: He was thrilled by the hope of possibly being rescued, but he was terrified by the fear that those on board might not see them and would pass them by.

"Hey, watch it, Gary!" shouted Jeff Hinshaw as he paddled

after his friend. "The boat's going to run right over you! It's going to hit you!"

"I don't care," Gary Marlar called back. "This guy's not going to get by us!" And he continued to flounder his sluggish way ahead through the choppy seas, directly into the path of the vessel now pounding its way toward him.

Unbeknownst to Gary Marlar and Jeff Hinshaw, the swift tidal currents had carried them some thirteen miles from shore. The tired crew of the fishing vessel *Ten Bears* had worked through a long journey of pulling and moving gear out on the Kodiak king crab grounds, and now they were fast asleep. Only her skipper remained awake and on watch. It would be another two-hour run into town. He'd take her the rest of the way himself.

Moments before, skipper Jim Batlien checked the vessel's course heading and went below into the ship's galley to pour himself a cup of coffee.

Never before, in all the thousands of hours he'd used the automatic steering system, had Batlien experienced any problems with it "whatsoever." He had only to set the ultradependable steering unit on a given course and away the ship would go.

But this night, something happened, for he had hardly reached the galley when the automatic steering system "went haywire," launching the *Ten Bears* on an entirely new heading. Standing in the galley, the sudden course change staggered Batlien. When he had regained his balance, he ran up the stairwell into the wheelhouse.

He couldn't believe it. The Iron Mike compass heading had leapt a full twenty degrees to starboard. The automatic pilot had failed him. He would have the system checked out once back in port. Then, as he leaned forward to correct the course heading, he spotted the flash of what looked to be a man in the water. The man was but a few yards away. He was waving frantically.

Gary Marlar slapped at the side of the *Ten Bears* with both hands. He would later remember the ship slowing, the light on the water, the rustle of crewmen scrambling above him, and the feel of a three-foot steel picking boom (a three-foot-long steel J-hook used to grab and haul king crab pots aboard) snagging him through the sea-bloated survival suit fabric near his knee. He would also remember being lofted into the black and windy night, and finally, the strong, excited hands yanking him aboard.

Gary Marlar and Jeff Hinshaw had been adrift at sea in the Gulf of Alaska for nearly fifteen hours when Jim Batlien lucked upon them. Jeff Hinshaw's survival suit had served him well, yet while he had arrived on the deck of the *Ten Bears* in fair condition, Gary Marlar was "clearly dying."

Marlar had no sooner touched down on deck than the crew of the *Ten Bears* went to work. They stripped Marlar naked, carried his blue corpselike body inside, and lifted him into a hot shower. But nothing seemed to help. After that, things begin to fade for him.

With the two men safely on board, Jim Batlien gunned his engine and headed for town. Then he sent out a Mayday call to the U.S. Coast Guard base in Kodiak, asking that someone meet them on the docks there when they landed.

It was nearly dawn when the *Ten Bears* finally pulled into

the port of Kodiak. Insisting that he could walk, Gary Marlar stood for a moment, then promply collapsed onto the dock. The crew of the *Ten Bears* gathered him up in their arms and rushed him up the docks to the waiting ambulance.

As he was wheeled into the hospital, he was greeted by none other none than Dr. James Halter—the very same doctor who, not two days before, had released him from that same hospital on the condition that he would go home and take care of himself.

"Gary Marlar!" he yelled when he first spied him lying on a table in the hospital's emergency room. "I thought I told you to go straight home and go to bed!" Marlar's skin was so blue "it looked almost black," recalls his wife. And as he lay on the hospital gurney, Gary Marlar lifted his weak head and flashed a glance in the direction of the doctor. "What can I say?" he mumbled, falling back in exhaustion.

In the grandiose thoughts that often accompany severe hypothermia, Gary Marlar thought he felt good enough to go out and celebrate. "I wanted to go out and get good and drunk," he recalls. In reality, Gary Marlar's body temperature was a mere 74°F. He would not be able to walk for two whole months.

Jim Batlien, the skipper who rescued the men from their fatal predicament, would never be able to explain the sudden mechanical malfunction on that dark October night. "That steering system had never given me a bit of trouble," he told Gary Marlar after the incident.

Several years later, Jim Batlien would die at sea in yet another mishap in the dangerous waters off Kodiak Island.

LOST AND ADRIFT

White, with green trim, the fishing vessel *Cloverleaf* stretched just sixty-three feet from bow to stern. Built originally in Astoria, Oregon, she was small by Alaskan standards—"just an old wooden whore" of a scow, Rick Laws recalled, when he and a friend, Dan Nelson (affectionately known as "Slippery Dan"), went to work rebuilding her. The overhaul quickly became a love-hate affair but with a tremendous amount of time, effort, and money, they were eventually able to restore her.

Having rejuvenated her into a first-class little shrimp boat, they fished the *Cloverleaf* hard. And in the years to come she proved to be a dependable and productive vessel, a midsized highliner of sorts.

Each winter, before the shrimp season began, Laws and his crew would use their trawl net to drag up tons of cod and other fish from the bottom and sell them as bait to tanner crab fishermen who worked the shores all along Kodiak Island. He and his two-man crew had managed to carve out a pretty good living for themselves.

Depending upon the season, those journeys took them over on the Alaska mainland, and even as far west as Dutch Harbor in the Aleutians. The Alaskan mainland had always been a vast wilderness country, and it was while traversing the hundreds and hundreds of miles of wilderness shores between Kodiak and Dutch Harbor that the *Cloverleaf* was occasionally pushed to the limits of her capacity. Many of the bays in which the *Cloverleaf* rendezvoused with the crab boats and delivered her fresh hold of hanging bait were uncharted, and remain uncharted to this day.

The *Cloverleaf* had a radar that worked "most of the time," and a good compass. But she had no loran, no running lights, and no auxiliary engine. In favorable weather, she topped out at six knots.

During the winter, travel along the Alaskan mainland often includes cruel and extended blasts of arctic weather. Heavy icing conditions are common. And then, with so little modern electronic navigational equipment on board, piloting the *Cloverleaf* often became a matter of guesswork, improvisation, and fly-by-the-seat-of-one's-pants navigation.

During the winter of 1980, Rick Laws faced just such a situation. He had received word that a fleet of crab boats was fishing for tanner crab in the waters around Sutwik Island, down off the Alaskan mainland. Normally, Laws knew, they wouldn't run that far. But there were a lot of crab boats fishing in the area, and they'd probably welcome his business. Once there, he knew they could anchor up on the lee side of the island and ride out even the worst of weather. The *Cloverleaf* had a nice load of bottom fish on board, so he thought, Well, let's take it to market. It would be a fine payday if everything went as planned.

It was dark when Laws motored past Cape Alitak on the south end of Kodiak Island and out onto the unpredictable waters of Shelikof Strait. The waves were moderate, the winds steady.

It was not until they had moved into that huge triangular body of sea stretching from Cape Alitak to the Semidid Islands and west to Sutwik Island that the situation grew extreme. In no time, it seemed, the wind stiffened sharply and tall seas arose. The *Cloverleaf* began to leap and roll. As they drew closer to the Alaskan mainland, the temperature dropped abruptly. And in the late-evening light, a heavy snowfall rolled in over them, shutting down visibility and leaving them in whiteout conditions. Now forty-knot winds blew a blizzard of snow in blinding slants across the heaving gray face of the sea.

When Rick Laws awoke to relieve Crewman Kim Handland, one of his two deckhands, from his watch, it was pitch black outside and the seas were "sloppy as all hell." There was no way to illuminate the darkness, as the *Cloverleaf* had no running lights. "Basically she just had a hull and one old screaming Jimmy," Laws recalls, referring to the vessel's diesel engine.

With darkness closing all around, Rick Laws considered the building sea and the blizzard of wind-whipped snow. Outside, the tall figures of the ocean waves, some fifteen feet high and more, began lumbering out of the gray-black night like mammoth shadows. They crashed repeatedly across the *Cloverleaf's* back deck, flooding around the wheelhouse itself and spilling over the side.

Laws came awake fast now. He gulped coffee and began

113

mentally updating his position. He had a good depth finder in working order and a complete array of marine maps of the area, so in theory he'd be able to use both in order to help him find his position. In addition, the boat's travel time, compass reading, and estimated speed could all be used to help fix the *Cloverleaf*'s position.

With the help of a good meter, a chart, a compass, and the moody output of a very old Decca radar system, Laws had always been able to find his way. He often dryly referred to it as his own style of "dead reckoning."

Laws remained alone at the helm through the long, battering night. Then at about dawn, he looked out on the back deck and noticed that part of a pile of net stored there was beginning to work itself loose.

"Hey, guys," he called to his crew in their bunks below. "You need to get up and tie down the net that came loose on the back deck."

As always, crewmen Wink Cissel and Kim Handland rose without complaint. When they had dressed, Laws slowed the *Cloverleaf* and turned her bow into the oncoming seas.

It was then that Laws sensed danger. Idling into the dark green slopes of the approaching waves the crew noticed it, too. Normally quick to respond, the *Cloverleaf* now felt slow and cumbersome. As he stood at the controls, Laws noticed with growing concern that each time he turned the wheel, the ship would respond only after a lengthy delay.

Her sluggish feel wasn't the only thing that concerned Laws and his crew. The *Cloverleaf* was squatting heavily in the water. And each time a wave broke onto her deck, the deck remained awash far too long. It was not draining at the rate it always had. The *Cloverleaf* moved now in slow side-to-side motions. It was clear that she was having trouble recovering.

Off and on all night, Laws had been chatting over the marine radio set with Bert Parker, owner and skipper of the ninety-one-foot crab boat *Amber Dawn*. Laws now made a hurried call.

"Bert, I don't know," he said in a worried voice. "This thing's feeling awful heavy and the decks aren't clearing very fast." He paused. "You better stay on the air, because this just doesn't feel right. The seas are breaking over my decks."

Bert Parker knew Laws was in trouble. He could hear the tone of alarm building in his voice, and he agreed to stand by on his radio.

Then, as Laws studied his radar screen, he thought he could make out land in the distance. A quick calculation told him he was looking at the tip of Sutwik Island. Believing that he was nearing the island, Laws once again called his friend.

"Bert, where's a good place around the island there to get out of this blow?"

Bert Parker was quick to respond. "Yah, just go ahead and come on in," he said encouragingly. "Stay about a mile off the tip of Sutwik and come on around and get in the lee of the island here. But I'd get your survival suits out and put them someplace where you can get to them quickly," he warned. "I'll be standing by here."

Laws turned and yelled to one of his crew members. "Wink! Run down in the engine room and check it out! See if there's any water in there!"

Wink Cissel leapt to the task and soon returned with good news. "Nope, engine room's dry! Everything's fine!"

Still Rick Laws didn't like what he saw outside. He rushed to call Parker back. "Bert, she just doesn't seem to be able to shake the water," he radioed. "I can't get her to come up. The deck won't clear and she feels real sluggish."

"Shit! You guys better get into your survival suits," warned Parker.

Turning to his crew, Laws yelled, "Everybody get your survival suits out and put them on!"

The suits were stowed close by in the galley for easy access. Laws remained at the wheel. In only the few seconds it took to shake the suits out of their bags, Laws felt himself losing control of the *Cloverleaf*. He grabbed the engine throttle and opened up the main engine, but the *Cloverleaf* refused to respond.

"Holy shit!" yelled Laws.

He grabbed the radio mike and called Bert Parker. "We're in trouble here! I believe we're about thirteen miles off the northern tip of Sutwik Island," he announced, yet even as he spoke, he knew this was only guesswork. The *Cloverleaf*'s pitiful radar system could not reach over to the Alaskan mainland. He couldn't even be sure it was Sutwik Island he was heading for. Perhaps it was the Semidis. But without another reference point, Laws had no way to get a triangulation and pinpoint his exact position.

"Yah, but where?" shot back Parker in frustration.

"I don't know!" shouted Laws as he struggled to pull on his survival suit. Just then, he felt the *Cloverleaf* begin to roll. Wink Cissel stuck his head in the back door and screamed, "Rick! Get out of there!"

Looking out the back door of the vessel he had come to depend upon for so long, Rick Laws took in what seemed an impossible vision. The stern of the *Cloverleaf* had sunk out of sight. Not a single deck plank remained visible. Her entire stern deck was submerged. And now a sprawling body of heaving seawater stretched out in every direction before lost to the slanting, gray-white oblivion of a blizzard snow.

Damn! thought Laws, fighting the urge to run. I've got to let someone know where we are!

Suddenly, the *Cloverleaf* rolled completely on her starboard side. As the vessel went over, Laws yelled into the mike; "Mayday! Mayday! Mayday! Hey, Bert! You guys better get out here, 'cause she's going down!"

In the next instant, Laws found himself tossed across the room. He landed on top of the radar set. In the inverted space of the wheelhouse, the port-side door hung above him. He still couldn't believe what was happening.

Holy shit! he said to himself. I gotta get the hell out of here!

He dropped the mike and began climbing up the face of the instrument panel. He ascended toward the rectangular space of gray light overhead. Once there, he pushed open the horizontal form of the wheelhouse door. Then he placed his hands on either side of the doorway, hoisting himself up through the rectangular space and out onto the side of the wheelhouse cabin, as if he were climbing into an attic space.

Back aboard the *Amber Dawn,* skipper Bert Parker could feel his blood pressure soar.

"Hold on! And stay together!" he shouted into his mike. "I'm coming full bore toward yah, Rick!"

When he received no reply, Parker rushed to notify the U.S. Coast Guard base in Kodiak.

"Mayday! Mayday!" he called over the far-reaching marine-band radio for the entire Alaskan fleet to hear. "This is the fishing vessel *Amber Dawn,* WTP-three nine zero five. I just got a Mayday from the fishing vessel *Cloverleaf!*" Then, locating the *Cloverleaf*'s last tentative position on his map, Parker quickly gave the Coast Guard his current loran readings and told them the area from which the *Cloverleaf* was believed to have last transmitted.

Rick Laws found himself standing on the side of the *Cloverleaf*'s wheelhouse. The circumstances he and his crew faced now seemed unreal, even fantastic. As he struggled to keep his balance, the ship he had rebuilt, piloted, and trusted for so long foundered beneath his feet.

And with nearly three-quarters of the vessel already under water, Laws rushed to finish pulling on his survival suit. Even as he zipped the final chin flap in place, seawater rose above his waist. Not five seconds later, he felt himself float free of the hull as the *Cloverleaf* sank out from under him.

The ship went down in an eerie silence. Small pockets of air boiled silently on the surface. For a space of time, Laws could see clearly into the depths. He could look directly down at the *Cloverleaf*. He would remember forever the vision of her sinking. As the boat descended farther into the deepwater currents below, she remained frozen in the exact horizontal position she had assumed on the surface: on her starboard side, her bow raised slightly.

The *Cloverleaf* descended with her interior lights on. Laws could see the entire boat at once now. And as she sank, he took in the strangeness of the image below. Her wheelhouse was still illuminated, and he could see directly into its submerged and flooded interior through her square-framed windows—windows that were now blazing with a warm, golden light.

Laws was amazed at how quickly her sixty-three-foot body sank and how minuscule she eventually became, free-falling into the depths far below. Long after he lost sight of the vessel itself, he could make out the tiny and yet distinct yellow squares of light still glowing against the greenish black back-drop of sea space.

"Hey, Laws!" shouted someone close by in the water. It was Kim Handland, the youngest crewman on board; his voice sounded shaky and disheartened. "I forgot my suit!"

The skipper's heart fell. The young deckhand had panicked and left his only hope for survival behind inside the boat.

"Hold on!" Laws hollered back.

Though only minutes had passed, he could see that Kim was already having difficulty staying afloat. The young man couldn't seem to catch his breath.

Laws swam over to Kim and grabbed him. "I've got you, man! I'll hold you," he assured his friend.

Yet secretly, Laws held out little hope. There's no way, he thought as he held on to the cold-ravaged crewman. There's no way I'm going to be able to get him out of this alive.

Laws knew there was an inflatable collar pillow built into the vest of his survival suit somewhere. But the two men were struggling mightily just to remain afloat, and Laws decided against trying to inflate it.

In but a few short minutes, Kim grew so numb in the 38°F. seas that he could no longer move his legs or his arms. And a howling wind of approximately fifty knots drenched the two with an icy, stinging spray.

Locked rigidly in his skipper's arms, Kim could only try to remain afloat as long as possible. He stared into Laws's face as they drifted down through the deep wave valleys and then over the tall wind-whipped crests.

Rick Laws's arms ached from the life-and-death effort to carry his friend. The aching in his upper body gradually went beyond fatigue, passing into a kind of pain-racked numbness. Now it took everything he had to hold on to his helpless companion and keep them both afloat. It amazed Laws how quickly immersion in the cold water had rendered the young man completely immobile. The wintery Gulf of Alaska water had "sucked the strength right out of him."

The last words Kim spoke were calm and matter-of-fact. "I can't move. I can't feel," he said with resignation.

Rick Laws looked into his face. "Don't worry, Kim. I'll hold you" he said.

But even as he spoke, Laws felt his grip beginning to fail him. He knew then he wouldn't be able to hold on to Kim much longer. "Oh dear God," he prayed. Kim Handland had worked for Laws since he had gotten out of the Coast Guard. He was a hard worker, a nice, quiet, and dependable youngster with as fine a combination of personality traits as one could hope to find in any man.

Near the end, Kim did not seem to be breathing. He uttered no words. His body drifted alongside Laws apparently void of feeling. And as they passed together through the icy flood of sea and spray, he grew completely rigid.

Laws could feel himself gasping for air as he struggled against exhaustion. He could see no sign of life coming from Kim—not a word, not a sound, not a movement. Nothing except, that is, for the young man's eyes. They remained wide open, fixed on Laws through the entire ordeal.

When an exceptionally large wave broke over the two, Kim's rigid body was torn from Rick Laws's weary grasp, and he immediately began to sink. Laws knew it would be impossible to dive in the survival suit he was wearing. Besides, he was far too fatigued even to attempt it. Laws could only watch as the body of his friend drifted toward the bottom, a thousand feet below. But the image that would remain forever etched in the skipper's mind was that of Kim's frozen gaze staring back at him as the young man slipped into the dark green depths below.

Now, as Rick Laws drifted over the steep and building storm waves, he was struck by a wrenching sense of loneliness. He'd fought hard to save the man's life, and the sudden loss left him dazed and confused.

Winded from the effort, stunned by the outcome, and chilled by the cold pressing in on him, Laws knew he had to rest. He squinted against the blinding sea spray and managed to locate the black inflator hose attached to his vest. He placed the hose end against his lips and blew. The vest billowed around his neck in the shape of a circular pillow. Even as he blew, he could feel his need to struggle easing as the buoyant force of the air bag began to lift and carry him.

Laws felt a degree of satisfaction at the result. He tightened the seal cap down over the end of the inflator hose. All right, he said to himself with renewed hope. This thing just may work. But no sooner had he leaned back to relax and gather himself than the entire inflator hose unit exploded from his vest. With a hiss, the defective unit launched itself into the air like a missile. The vest around his neck deflated suddenly and completely.

Laws realized angrily that now all he had left was his own power: He would have to keep treading water if he was to remain afloat—and alive.

In launching his search from aboard the *Amber Dawn,* skipper Bert Parker went over and over his final conversations with Rick Laws. When Laws had last called him, he had reported that he could see a landmass he believed to be Sutwik Island.

Laws had sounded "reasonably close" to Sutwik, Parker figured, but there was no way to know for sure.

Five crab boats immediately joined Parker in his search effort, which would continue throughout that first day and on into the night. "It was blowing sixty out, with big running seas," Parker recalls. "There were five or six boats involved in the search. And, oh, man, did we ever get our butts kicked!"

Parker felt emptied by the news of his missing friend and crew, and he stood at the wheel through the long hours ahead, refusing all offers to spell him. Crowded into the wheelhouse of the *Amber Dawn,* every crewman on board scanned the sea for signs of wreckage or life. But throughout the long day and well into the night, a blinding curtain of blizzard snow swirled across the watery landscape of broad, rugged waves, making visibility all but impossible and diminishing the chances of locating anyone still afloat.

That night, the outdoor temperatures fell well below freezing and ice began to coat the deck and superstructure of the *Amber Dawn* and the other crab boats involved in the search. But Parker and his fellow skippers continued their efforts.

Parker pounded his way over and through the unrelenting storm waves for more than twenty-four hours. His weary eyes scrutinized the face of each new wave as it appeared out of the gray-black depth of blizzard storm. He searched the offshore waters around Sutwik Island for a distance of more than fifteen miles, then hunted downwind from the nearest edge of the island for another twenty miles. But he found nothing.

Now as he drifted and struggled to remain afloat, Rick Laws found himself replaying the last few hours in his mind. He could recall no major mistake, nothing he had done wrong that could have led to this. In prep for running, the crew of the *Cloverleaf* had battened down her hatches, thrown a tarp over her back tank, tied and retied everything that could move in place, and lashed it all down for traveling. Then they began taking turns at the wheel. As always, Laws rotated the men: Too long alone at the helm and any deckhand might fall asleep on duty. (On the other hand, too short a wheel watch would mean that neither of the other two crewmen could get any sleep.)

Rick Laws was six one, weighed a solid 190 pounds, and sported a full black beard and a head of curly black hair. A natural athlete, he had competed as a long-distance runner.

Laws and crewman Wink Cissel had attended rival high schools down in the Sonoma wine country of northern California. At six four and 175 pounds, Wink Cissel was tall and lanky and as lean as iron. Laws both loved and respected the man. "He could work forever at his own pace," recalls Laws. "And he was no complainer." Some guys have a natural talent for telling a story. Sit old Wink down on the waterfront sipping a couple of brewskies, and pretty soon there'd be six or eight guys standing around listening to him spin a yarn. "He was a natural for that. He could hold court just about anywhere."

Rick Laws had originally come to Alaska in 1975 to photograph Wink Cissel's wedding, then ended up staying in Kodiak when he got a job sharpening J-hooks aboard an old halibut schooner.

Laws built a small cabin on a corner of Wink Cissel's property and lived there while he worked with Cissel building a house overlooking Monaska Bay. Each morning before the work began, Rick Laws would hike up the hill to have coffee and breakfast with Cissel and his wife, Janet, and their family.

And Wink's kids always greeted him as part of the family, calling out, "Uncle Rick! It's Uncle Rick!" whenever he appeared.

After the *Cloverleaf* sank, Laws had briefly caught sight of Wink "three or four wave sets" in the distance. Laws had tried to scream, "Are you okay?" But the waves were tall and the wave valleys wide "and the wind was just howling." Then Laws caught sight of Cissel again as he drifted over the top of a wave. He had his hood up and his flotation device inflated. "He was lying back in the water like a man riding an easy chair," Laws recalls.

Rick waved. Then, as Wink drifted over the top of another large wave, he signaled Laws back with an encouraging "thumbs-up" sign. Rick Laws never saw his friend alive again.

Up until that time, Laws had been concerned only with the welfare of his friend. He hadn't had much time to worry about himself. He'd been safely locked in that focus, and he had found comfort there. But Kim had died. And his good friend Wink had drifted away into the blizzard and seas. And when his own inflatable survival suit vest failed him, Law's mind raced to deal with a new and monstrous reality.

Holy God, I'm totally alone out here! he thought. It was a sickening and frightful aloneness, the "most alone" he could ever imagine one could be. He was plagued by bouts of sadness over his friends, by worry and thoughts of doom. He was forced to admit to himself, Hey, I may not get out of this damned situation alive!

The thought of never seeing his mother and father again wore at him. His mind flashed to his father. He could picture him sitting on the back deck on a hot and sunny California afternoon with his shirt off, holding a cold bottle of Michelob in his hand and spouting funny stories to the men working in the stables.

Then, in the darkness, Laws somehow managed to spy the shadow of a log drifting past. New life shot through him. If I can get on that, I'll make it, he thought. Though exhausted, he "swam and swam and struggled like hell to get over the seas" to the log. But when he caught up with it, he was discouraged to find it was almost no help at all. Every time he tried to lean on the log or climb atop it, the log rolled. He had to keep treading water even to hold on to one end of it.

So engrossed was Rick Laws in his minute-to-minute struggle to survive that he had lost all sense of time. The short, dull gray durations of winter had come and gone, daylight and now darkness surrounded him. It was near midnight (more than sixteen hours after the *Cloverleaf* had sunk), that Laws felt the most hopeless. And as the cold and waves continued to bombard him, he felt almost undone by the savage bouts of hunger, thirst, and exhaustion that hounded him.

At one point during the night, a sooty sheerwater bird landed on the opposite end of the log from him. You know, thought Laws, if I can just get ahold of that bugger, I'm going to rip its head off, eat its heart, and drink its blood. The dark outline of the bird resting on the log seemed to indicate it was sleeping. In the late-night darkness, with the constant commotion of water lapping over the log, Laws figured he had a good chance of sneaking up on the bird.

Carefully, Laws made his way along the back of the log. He had maneuvered only halfway down its length when the bird leapt into the air and disappeared. Laws cringed, discouraged and almost beyond hope. Oh man, he thought dejectedly, that bird might have made the difference.

Ever since Kim had died, many of the beliefs to which Rick

Laws had always held fast had undergone an onslaught of change. Confronted with a life-and-death need to adapt, Laws found himself abandoning the old for the new, shedding past alliances suddenly and completely. And in the long, cold, body-numbing hours of struggle and suffering ahead, Laws would often be struck by a sense of wonder at the exceptional nature of thought and emotion that came to him.

Already, his mind had undergone several transformations. Getting through the initial denial was one: He had been unable to accept what was happening. His first thought when he had hit the water and was left to drift in that wilderness of surrounding sea was Damn, I'm sure going to miss my truck! It wasn't until he saw Kim Handland drifting toward the bottom that he came to terms with the truth. In accepting the death of his crewmate, he had also forced himself to acknowledge the possibility that he, too, might die. And in doing so, he had faced the primitive and brutal nature of his struggle. He saw then that he was utterly alone. Fear was another stage: Finding himself adrift in the night in a haunting black chamber of shadows and sea was a frightful experience. Yet during that time, Rick Laws came to a gritty decision: He would keep himself afloat and do battle with whatever came his way. He would assume the entire responsibility for his survival.

It was fortunate that Laws came so quickly to such a conclusion. For all that night, he fought against the loss of strength as the icy seawater enveloped him and tried to sap his body heat. His belly gnawed with insidious bouts of hunger and thirst. Laws felt himself pass through entirely new thresholds: The racking pain and numbing cold were still there, but once he'd made his decision to fight they paled alongside his fresh desire to remain upright and alive.

Ideally, when a boat sank, the crewmen on board zipped

up their suits, and someone passing nearby came immediately to rescue them. But Laws knew it seldom happened that way. With the huge expanse of Alaska's wilderness waters totaling well over a million square miles, there was always the possibility that no one would find him. Not ever. The older-model survival suit Laws wore wasn't designed for long periods in the water. In fact, his suit was leaking all over. Small leaks had started in the crotch area and the wrists. And he could feel icy rivelets of seawater seeping in around the rim of the hood encircling his face.

That night, Laws spotted the mast lights of numerous crab boats searching back and forth across the seas all around him. He watched them draw "closer and closer" and thought to himself, There's a chance! At least somebody knows I'm out here.

Laws could see their search lights rising and falling off in the darkness as their boats pounded their way through the lumbering seas. The boats zigzagged sharply back and forth across the water. He could tell they were working a grid pattern.

Suddenly, Laws remembered a small reflective strip sewn into the sleeve of his survival suit. And as the ships drew ever closer, he raised his arm and began to wave it wildly through the air, back and forth overhead.

"Oh man—you guys! Come on! Come on!" he pleaded, yelling into the wet night air.

Each time he drifted over one of the fifteen-foot wave crests, Laws would kick hard, rise far into the night, and wave his reflective strip.

Not long after that, he watched in utter disbelief as the entire crab fleet shut off its mast lights, turned around, and steamed away into the night.

"Oh, God! I'm not going to make it! I'm a dead man!" Laws shouted after them.

He knew now the odds of surviving his ordeal had just dropped to "near zero." Regardless, he refused to quit. The thought of breathing seawater filled him with a sickening fear, and once again he renewed his vows to himself. He resolved to fight to the end, to do everything he could to remain alive, to keep breathing air for as long as he was physically able. He would never give up!

It was perhaps 3:00 A.M. that night when Rick Laws had the vision. It came in the form of a "huge black sailing ship." The gleaming vessel sailed out of the obscurity of the night as if on a mission, and cut a direct course toward Laws. It was an eighteenth-century sailing ship. But the strangest thing about it was her color. Her hull and her billowing sails, too, were all jet black, an oily black. The ship bore down on Laws, glistening in the night as she came. And every part of her shone with the rich blue-black gleam of gunmetal.

"It's real," said Laws, astonished.

As the ship drew near, the notion struck him that if he could climb aboard the vessel, he'd know "complete comfort." Laws felt he was being given a choice. He could continue on with his painful struggle, or he could die. He viewed the crucial decision as a "crossover."

At that moment, an electrical charge shot through the length of Laws's body. The hackles on his back went up, and without hesitation, he spoke aloud to the vision. "No! Wait a minute! I haven't had my last Heineken yet! I'm not done! I may not have much hope of getting out here, but I'm not getting on your ship! And I'm not breathing seawater."

No sooner had he spoken than the phantom ghost ship turned and sailed away, vanishing into the night.

Rick Laws could feel his temperature dropping. He knew that he was dying. Now as he passed through the ever-increasing thresholds of pain, his mind took a course all its own.

His suffering brought one fact home to him: "When you're so cold you can no longer think, and you pass through all the emotions and all the pain, there's always someone still home in there—in your heart cave."

He was still paddling when the first gray light of dawn drove away the darkness. He was comforted, for a moment. But the hope he had at first felt rising was quickly overwhelmed by another harsh fact: Laws realized that all the reserve of strength he had called upon through a day, a night, and into yet another day, had finally been exhausted. Now nothing was left. His arms and legs were both failing him. He was nearing the point of complete and catatonic exhaustion. Rick Laws had no more resources left to keep himself from rolling face-down and breathing seawater. Short of a miracle, he was certain he wasn't going to make it. Unwilling to accept the terminal end to all his efforts, Laws rolled slowly over onto his back. Then, looking up at the overcast skies, he called out, "God! If you're there—really there—I need a miracle."

Within minutes, the "wind stopped, the sun came out, and the seas came down." Every hair on Laws's body stood up. At that moment, he knew his cry had been answered. He "knew in his heart" that he was going to be rescued. For the first time in his life, he believed he knew what it was to have faith.

Not more than an hour had passed when a C-130 SAR plane

from the U.S. Coast Guard base in Kodiak flew onto the scene. Laws heard the plane approach in the distance, and he smiled inwardly to himself. Yep! Here it comes! I'll be damned. He's going to find me. I'm going to make it. I'm going to make it!

When he first saw the plane, it was moving low across the water, well off to the side of him. Laws lay back then and slowly began fanning his arms and legs at his sides, like a child making angel wings in freshly fallen snow.

Seated in front of a spotter's window inside the C-130 was PO Brian Blue. He was looking down and across the water at the time, and he spied the drifting hulk of a log. With the plane moving across the face of the sea at 210 miles per hour at an altitude of five hundred feet, he found he was having difficulty estimating the true dimensions of what he saw below. Almost simultaneously, however, he spotted what looked very much like an orange starfish drifting next to the log. But there was something odd about this starfish. It had only four tentacles. Not until he saw those same tentacles move did he realize he'd spotted the body of a man adrift in a survival suit. The slow-moving survivor was apparently attempting to signal him.

"Hey!" he shouted to the crew inside the plane. "I think I see one of the men down there!"

As Laws watched, the C-130 banked sharply and, with engines roaring, flew back toward him. The aircraft swung in low across the wave tops and over him. Laws could only lie back and wave tiredly. The plane swooped in repeatedly over him and continued to circle, dropping numerous colored flares and dye canisters in the area to mark his position. Once, Laws watched the C-130 until it had almost disappeared into the distance. It had only just begun to bank and return again when it suddenly veered back on course and continued out of sight.

He would later learn that at that moment the pilot of the C-130 had spotted the body of Wink Cissel drifting in his brilliant orange survival suit on the surface. Though the vest of his suit had remained fully inflated, keeping him afloat, Wink had nevertheless died of hypothermia during the night.

After more than a day of fruitless searching for his missing friend, Bert Parker and his fatigued crew aboard the *Amber Dawn* were forced to conclude that Rick Laws and the men of the *Cloverleaf* had most likely perished.

Parker was certain the *Cloverleaf* had sunk. And even if those terrified and disoriented crewmen had managed to get into their survival suits, no human being in the history of Alaska had ever survived so long adrift in such icy seas. The manufacturers of these suits claimed protection from the cold and hypothermia for only three to five hours.

Parker was deeply saddened. He had just headed in to drop anchor and get some much-needed rest when something strange happened: an abrupt change in the weather. The forecast had called for gale warnings in the area, but suddenly the wind died down and the waves eased noticeably. Then the clouds parted and the sun came out. It was a dramatic and unexpected turn of events.

The U.S. Coast Guard quickly took advantage of the improved visibility. Minutes later, a C-130 SAR plane flying some forty miles from Sutwik Island (tens of miles outside the area where Bert Parker and the other crab boats had been searching) radioed that they had spotted what they believed to be a crewman floating below. And the crewman was apparently still alive! Parker and his dog-tired crew were jubilant. The crewman would turn out to be none other than Parker's good friend Rick Laws — in critical condition, but alive.

Vern Hall, skipper of the 121-foot crab boat *Rondys,* was the first to reach Laws amid the wafting clouds of flare smoke.

Despite the optimistic circumstances, Rick Laws was now nearly unconscious from the cold. Severely hypothermic, he could no longer make sense of anything. He could tell the *Rondys* crew was yelling to him; perhaps they wanted him to grab the line drifting nearby. But all his brain could think was, My God, that boat's big!

Vern Hall could see that Laws was too weak to grab hold of the line thrown to him—too disoriented, for that matter, even to locate the line directly beside him in the water. It was *Rondys* crewman Terry Sampson who quickly donned his survival suit, leapt overboard, and with rope in hand, swam out to Laws, trailing the line behind him.

Sampson wrapped the line of the buoy hook under Laws's arms, then signaled the crew to pull him in. Laws could hear the whine of the hydraulic king crab block as he was dragged through the water. In quick succession, he found himself yanked alongside the huge steel hull of the *Rondys,* craned into the air, and hoisted over the tall handrail and aboard.

"Vern Hall did a good job, I know that," Rick Laws later said appreciatively.

As he lay on the deck of the *Rondys,* Rick Laws discovered that his "body wasn't working," as he put it. Not only was he unfit to walk; he couldn't even manage to crawl. He could hear the remote voices of the *Rondys* crewmen, Terry Sampson and David Capri, as they rushed to save him.

They wasted little time. One moment, he was lying in a helpless heap on the wooden deck; the next, they were stripping his survival suit off him. Then he felt the comforting grip of many arms hoisting him from the deck. They ran carrying him into the wheelhouse, through the galley, and deposited him into the nearest stateroom, in the soft, warm, dry space of one of their own bunkbeds.

When skipper Vern Hall came into the room, he took one look at Rick Laws's blue lips and sunken eyes and knew the man was close to death. Hall asked for a volunteer to help warm Laws and stop the hypothermic fall of his body temperature. "One of you guys strip down and crawl into the bunk there with that man," he said.

Terry Sampson volunteered again. He quickly stripped, then jumped into the bunk with Laws; the *Rondys* crew immediately rolled both men in multiple layers of wool blankets.

At first, Rick Laws was embarrassed by this intimate contact with another man. He knew he was cold, but he hadn't grasped how thoroughly chilled he'd become until he felt the amazing transfer of body heat coming from Sampson.

"I could literally feel the heat from that other guy's body just pouring into me!" he recalls gratefully. "The heat coming from that man went all the way through my entire body. I believe he saved my life."

When he awoke later, his body temperature was rising. As the *Rondys* crew kept close check on his tenuous condition, Laws lay back in the soft comfort of the bunk and thought back over the incredible adventure he'd just been through. He'd survived twenty-seven hours in the deadly cold Gulf of Alaska sea—longer than any other man in history.

OVER AND OUT

The fishing vessel *Angela Marie* was tied in a snug dockside berth in the fishing port of Pelican, Alaska, when early on a cold and blustery February morning in 1994 skipper Charlie Christiansen ordered his crew of four to throw off their dock lines and brace for a rocky ride ahead. Crewmen Kurt Kivisto, John Breezeman, Erik Kegal, and Alann Erickson leapt to the task.

The first day of the southeast Alaska tanner crab season was close at hand, and Christiansen, the forty-four-year-old skipper from Petersburg, Alaska, was determined to reach Idaho Inlet for the opening gun. Standing at the helm of the historic old fifty-four-foot Alaskan seine boat, Christiansen idled her along the waterfront and guided her smoothly from the harbor.

Charlie Christiansen had been fishing in the area ever since he was a boy. Born and raised in the fishing village of

Petersburg, he'd purchased his first commercial fishing boat from an old Norwegian in the area. Christiansen was only nineteen at the time.

Now, with a quarter of a century of successful fishing behind him and his reputation solidly established, Christiansen felt comfortable running the *Angela Marie*. He considered her a "fine sea boat," and though the temperature of the knifing wind outside was well below freezing, it was nothing he, his crew, and the *Angela Marie* herself hadn't faced many times before.

This voyage, however, would be different. For neither he nor his crew — nor the *Angela Marie* herself — would ever reach Idaho Inlet.

A lot of rich history surrounded the *Angela Marie*. She was a Hansen-built wooden seiner, constructed on the waterfront in Seattle in the late 1950s. At the time, she was one of the larger fishing boats in southeast Alaska. Many looked upon her as the queen of the fleet. She was versatile, and after thirty years of plying these waters, she was still serving as a crab boat and long-liner, in addition to seining throughout the traditional summer-long salmon season.

Even now, as the *Angela Marie* moved out from the partial shelter of Inian Pass and into Lisianski Strait, she was packing a full ten-thousand-pound workload of some thirty-five pyramid crab pots — the pots had been stacked on her back deck and were secured with hundreds of feet of crab-pot line tied in the form of a Spanish windlass knot.

Christiansen had planned to maneuver a course through Inian Pass, scurry across the width of Cross Sound, and then muscle his way up the more exposed waters of Lisianski Strait to one of his favorite fishing spots. Christiansen had tried to make the passage only the night before, but they had no sooner left the snug protection of Pelican when they found themselves faced with fierce winds and forbidding seas, so they quickly returned to the security of Pelican. The next day with the new season close at hand, Christiansen felt compelled to go.

But as soon as he entered Lisianski Strait, a close-cropped series of pounding seas arose, and he was greeted by winds of fifty to sixty miles per hour that howled determinedly over the face of the sea. Shortly, the winds began gusting to between seventy and eighty miles per hour, with cold williwaw blasts exploding down out of the snow-covered mountain passes at more than one hundred miles per hour. The wintery blasts of arctic air sent the crests of the waves blowing across the face of a rugged and building sea in a feathery mist.

Then, near the entrance buoy to Lisianski Strait, they encountered a seemingly unending series of tall and formidable breakers. The waves were twelve to fifteen feet high, recalled one deckhand; such an onslaught would have been daunting even in a modern one-hundred-foot steel Bering Sea king crab ship, but it was downright perilous in a heavily laden, wooden seiner. Worst of all was the wave length—the close proximity with which those steep slopes of ocean water struck them; this proved the *Angela Marie's* undoing.

They passed the corner buoy and pounded their way around a particularly craggy area with tall columns of rock rising up along the cliff-lined shore. Fishermen refer to it as Column Point. Now each time the ship lurched over a wave crest, she would pitch forward, plunging into the trough ahead, and her bow would scoop water, disappearing into the steep greenish black slope of the next fast-approaching wave.

It was then that Charlie Christiansen began to feel a change in the way the *Angela Marie* was handling. Not long after he'd "turned the corner," the *Angela Marie* "plowed into a truly big one." The body of the wave drove straight into the boat. A wall of freezing spray broke over her entire length. The impact brought her almost to a standstill.

Heavily laden with king-crab gear, buffeted by hurricane-force winds, and burdened by an untold tonnage of seawater now flooding across both bow and stern, the *Angela Marie* struggled to remain upright.

Something is grossly wrong here, thought Christiansen. Either the lazaret is somehow filling up with seawater, or—his mind raced—the boat is going down.

Their only chance was to take the ten-thousand-pound stack of steel crab pots sitting on the *Angela Marie's* back deck and dump the whole thing overboard.

Deckhands Erickson, Kivisto, Kegal, and Breezeman, all of Petersburg, were seated at the galley table when the boat hit a crushing wave. There was a pause, then an "eerie silence."

Then they heard their skipper yell to them: "Something's wrong here! I'm losing control of the boat! Get out on the back deck there and start dumping pots over the side!"

Kurt Kivisto and his crewmates ran back, threw open the galley door, and peered outside. They had piled the deck high with three individual stacks of crab pots. Miles of buoy line had been wedged in between. Yet the entire area was now buried between surging currents of seawater. The only thing that remained visible on their back deck was a lone tentacle of floating buoy line trailing away from them. The deck was listing to starboard. A wild sea swept repeatedly over the deck. Ocean water lapped at their feet.

"Holy Christ, this isn't right!" said Kivisto aloud. The crewmen joined him as he rushed back inside and slammed the galley door behind him.

Christiansen decided to try to turn around and run with

the overpowering seas; he spun the large spoked steering wheel hard to port. The *Angela Marie* hesitated, then began to swing around. But she had managed to complete only a portion of the turn when a freak wave drove into her. The rogue wave seemed to materialize out of nowhere. It struck the sluggish form of the *Angela Marie* as she lay broadside in the trough. The wave exploded across the vessel's entire length, swamping her stern and nearly rolling her over.

Impeded by the leaden weight of a flooded stern, Christiansen somehow managed to complete the turn. Then he gunned the ship's engine in an attempt to accelerate ahead and outrun the pursuing seas. But, burdened as she was, the *Angela Marie* could not comply. Now wave after wave began to pound in on her stern, staggering the foundering vessel again and again and threatening to sink her on the spot. "They just started freight-training us, one after the other," recalls one deckhand. "And the entire boat began to shudder like she was going to tear herself in half."

"This is it!" cried Christiansen aloud.

Then, without warning, the *Angela Marie* began to list heavily over onto her starboard side, and the crew knew "there was no doubt about it—she was going over."

Christiansen knew it was a desperate situation. The *Angela Marie's* stern had already slipped under. Much graver, however, was her list. She was leaning hard to starboard now and seemed determined to roll completely over at any moment.

Charlie Christiansen knew that both the wheelhouse and fo'c'sle exit were mounted on the starboard side. If the *Angela Marie* was allowed to roll in that direction, the only escape routes for him and his men would be cut off. Anyone caught inside would almost certainly be doomed.

Without hesitation, then, Christiansen cranked the ship's

giant wooden steering wheel hard to starboard and "fire-walled it," gunning the engine. The *Angela Marie*'s engine responded for the last time. Christiansen and his crew felt the balance of the struggling vessel shift. Slowly, the wheelhouse rose upright, but it stood there only momentarily before continuing on over and completing its 180-degree swing. Then the *Angela Marie* fell over on her port side, and lay there drifting in the water like a crippled duck.

Inside, deckhands Kivisto, Erickson, Breezeman, and Kegal waited for the ship to right herself. In the past, they had ridden the vessel through many a similar blast of stormy sea. But now the *Angela Marie* remained stalled on her side, completely exposed to the oncoming seas. And a ceaseless battery of ocean breakers continued to pound her.

Kurt Kivisto's mind raced. His mind leapfrogged to other times aboard ships he'd crewed in Alaska. He recalled a collision in the fog while under way near Petersburg, and another harrowing situation up in the Bering Sea in the middle of the night. This would be his third. "Not again!" he said aloud.

As the ship rolled, those standing inside the upended space of the ship's galley suddenly found themselves dodging plates and silverware as "everything not tied down went flying." It seemed that their whole world was coming apart. Suddenly, the large microwave oven broke loose from overhead. It free-fell through the space of the inverted room; the full weight of it slammed into the wall between two of the crewmen with a crushing force, demolishing itself on impact.

Kurt Kivisto had a terrifying feeling that the boat was going to roll the rest of the way over and trap them all inside. "The hell with this!" he announced suddenly.

The four crewmen turned to run. Kurt Kivisto and Alann Erickson slammed into each other. Alann's glasses went flying.

He paused and searched frantically. Unable to locate his glasses, Erickson raced to catch his crewmates.

Crossing their skipper's stateroom wall as if it were the floor, Kivisto, Erickson, Breezeman, and Kegal made their way through the interior of the *Angela Marie* and climbed outside through the vertical overhead opening of the wheelhouse door. Then they stood in the bitterly cold wind, balancing themselves on the pitching side of the wheelhouse now underfoot.

Skipper Charlie Christiansen knew the battle to save the *Angela Marie* was over. The storage box mounted to the starboard side of the wheelhouse held the ship's most precious commodity: the extra survival suits. With the *Angela Marie* sinking fast, he knew he had to get to those suits; he couldn't take the time to make a Mayday call. There wasn't time to do both. If the ship went down carrying their survival suits, they would all die.

Christiansen climbed out through the wheelhouse door, up and over the side of the wheelhouse, and then lowered himself down onto the steep vertical slope of the roof.

The box containing the survival suits was nearly submerged when he got there. He clung to the side of the wheelhouse with one hand and lowered himself down onto the storage box. With his free hand, he flung open the lid and yelled above the wind for help. A moment later, the face of Kurt Kivisto appeared above him; without a word, Christiansen began passing up survival suits as fast as he could toss them.

Two of the suits looked fine. But several turned out to be children's suits, and it soon became apparent that there wouldn't be enough adult-size suits to go around.

"Where are the rest of the survival suits?" yelled Christiansen.

"They're all inside!" Kivisto replied.

In the panic and terror of being trapped in the overturned vessel, the deckhands had escaped the sinking ship—but without their survival suits.

Then a voice began to echo up through a nearby breather vent. "Hey, you guys! I'm coming! I'm coming!"

Then a head poked up through the horizontal slot of the wheelhouse escape hatch; the welcome sight was deckhand Erik Kegal. As Kegal's head emerged, Erickson yelled, "No! No! Get our suits! They're in the galley, right below you there!" Kegal disappeared back into the hole. When he returned, he handed up the three missing survival suits.

As the ship's designated radioman, Alann Erickson knew it was his duty to make sure a message was sent. Terrified that the ship would capsize, entombing him inside, he nevertheless lowered himself back down into the sinking ship; he climbed forward into the wheelhouse, grabbed the radio microphone, turned up the volume, and began sending out a call for help.

"Mayday! Mayday! This is the *Angela Marie!* The *Angela Marie!* We're in Lisianski Strait! We're lying on our side and we're going down! Will somebody please come help us? Please!"

With the vessel sinking out from under him, Alann Erickson dropped the mike and climbed frantically up through the rigging of the foundering ship. When he returned to the rest of his crewmates, they were still balancing on the starboard side of the wheelhouse. They were squatting there in the biting cold, balancing "like four seagulls on a log" as they struggled to pull on their survival suits.

As the *Angela Marie* continued to sink, her wheelhouse slumped slowly over, the angle of its repose growing sharper with each passing moment.

It was a gloomy sight. The crew had climbed free of the

rising Alaskan 40°F. seas, but they knew that at best they had secured for themselves only a temporary respite.

The biting edge of the wild, wintry gusts continued to drive a close-cropped series of pounding waves onto the vessel's stern. With the *Angela Marie* lying on her side, awash in the heave and roll of the seas crashing in against her, the explosions of freezing ocean spray drenched them all.

Far across the rugged waters, a rich green hemlock forest was visible, hugging an unattainable shore. The mountains rose almost from the waterline, verdant forests giving way abruptly to steep alpine meadowlands. Brilliant green throughout the short summer season, these highland slopes lay dormant and brown throughout the long Alaskan winters. High above them rose the sharp-crested mountaintops themselves, surrounded by a glacial landscape of snow and ice. Bone-chilling blasts of arctic wind racing out of mountain passes now dusted the skies with feathery swirls of powdered snow.

As the *Angela Marie* sank from under them most of her crew began to flee overboard one by one.

Yet even as the others leapt overboard, Kurt Kivisto remained. He kept saying to himself, It's not going to sink! It's not going to sink! I'm going to stay here until the bitter end!

As the ship slipped farther into the pounding seas, smoke began to belch out of the wheelhouse door. Then another wave broke over the crippled hull of the *Angela Marie* and the huddled figure of Kurt Kivisto. The flooding force of the wave caught Kivisto unawares, and it very nearly washed him into the doorway opening and back down into the sinking vessel.

Drenched and stunned by the sudden force of the icy seas, Kivisto, too, leapt overboard. He plunged in feet first, submerged briefly, and was greeted by the immediate and unwelcome shock of ice water flooding in around the neck of his suit. Unbeknownst to him, he had failed to close the last few inches of his front zipper, and now his suit was filling with a breath-jolting river of Gulf of Alaska water.

Unable to figure out what was happening, or why, Kivisto began struggling to remain afloat. There hadn't been time to launch the life raft, so he began searching the surrounding area for "something to hold on to."

Then he remembered the inflator hose. It was built into his suit and it led to his flotation vest. If he could only find it, he could blow up his vest manually. He searched frantically for the hose but failed to locate it. In frustration and fear, Kivisto began fighting against the suffocating flood of seawater, but that only seemed to make things worse. The more he struggled, the harder the waves seemed to break over him, and each time another icy jolt of seawater poured into his suit.

As he sank farther into the chopping seas, wave after wave broke over his head, now barely visible above the surface. And Kivisto was forced to face the fact that he might very well drown. "Man, this isn't working! This is just making things worse!" he said to himself.

Kurt thought of his family. He had three fine children, two girls and a boy. His wife, Sharon, was pregnant with their fourth child. Thoughts of his family lifted his sagging spirits and renewed his determination. And Kivisto made up his mind. He would do whatever he had to do to survive and go home again. Exhausted from the struggle, and winded from having to hold his breath each time a new wave crashed over him, Kivisto decided, to "relax and just go with the flow and see what happen." Almost immediately, things seemed to improve.

Suddenly, out of the corner of his eye, he caught sight of a cluster of buoy bags floating past. He recognized them as the crab-pot buoys that had been tied to the railing of their boat, and he knew they must have broken free when the ship sank. As the cluster drifted past, Kivisto stroked hard for them and managed to get a viselike grip on the line leading to one buoy.

"Thank you, God!" he cried aloud.

When he looked up, he spied Erik Kegal only a few feet away. Kegal was floating in the middle of the protective cluster of buoys like a sea otter lolling in a bed of kelp. Kivisto yelled to be heard above the rushing wind and broken slap of the water. "Thanks for coming and getting me, Erik!"

Erik Kegal yelled back. "Well, thanks for coming and getting *me!*"

Just then, Kivisto heard a fearful howl. He spun around in time to take in a peculiar sight: A noisy wave was breaking directly toward them from perhaps fifty feet away. And riding along atop the crest of that tall white breaker was none other than the wide-eyed figure of fellow crewman Alann Erickson.

He rode the wave like a bodysurfer, and he would have been washed well beyond the cluster of buoy bags had he not managed to snare one of them as he passed, freeing himself of the wild ride.

As Kurt Kivisto's suit continued to leak, he clung tightly to the buoy bags. When Alann Erickson arrived, he rushed to confide in him. "Damn, Alann, I thought I was going to buy it out there. I couldn't stay afloat. Seawater is still flooding into my suit."

Erickson paddled closer and quickly spotted the problem. "Oh hell! Your zipper's not pulled all the way up!" Kivisto felt foolish for having failed to get his hood snugly in place before tucking his arms inside. He knew that once your hands were inside the flipperlike gloves of most survival suits, it was almost impossible to get the essential hood area around the face sealed tightly in place.

Erickson searched for his crewmate's inflator hose and soon discovered that it had somehow wound itself around Kivisto's back, making it impossible for Kivisto to reach it. "Your inflator tube is behind you!" Alann yelled, pulling it free.

Kivisto seized the hose and began blowing long lungfuls of air into the tube, quickly inflating his vest. The newfound buoyancy was a tremendous relief.

Now, as the three crewmen drifted, they pondered the fate of their skipper, Charlie Christiansen, and their crewmate John Breezeman. "Guys," said Alann Erickson, "I don't know if Charlie made it. The last I saw of him, his survival suit was still unzipped and he was taking in water. And I'm almost positive John is dead, because he didn't get his suit on at all before he got in the water." He paused. "I guess the best thing we can do right now is to say a prayer for those guys—and for ourselves."

As they rose and fell over the driving seas, the wind howled through the bright orange plastic buoys. The three remaining crewmen huddled close, bowed their heads, and prayed aloud. "God, please help us to make it through this situation. Please let it be that someone heard our distress call. Please help John. And please help Charlie. The last any of us saw them, it didn't look good. Please let them be okay. Amen."

Moments later, another tall comber broke down upon them. Erickson saw it first. "Hang on! Big wave coming!" he warned. And the men turned their faces away and braced for the icy wash as the breaker roared over them.

*   *   *

Charlie Christiansen was still clinging to the side of the *Angela Marie*'s wheelhouse when he caught the harried flight of the newest crew member. There was no time to warn him as the panicked form of John Breezeman fled out the back door. Survival suit in hand, he literally ran into the ocean as the back deck sank out from underneath him.

Immediately, the tidal currents began pulling at him. Dressed only in his cotton work clothes, Breezeman was quickly carried away from the ship. He struggled mightily to pull his survival suit from its bag.

Charlie Christiansen climbed to the uppermost portion of the wheelhouse, then hurried to pull on his own survival suit, making a point to "keep an eye" on the struggling crewman. Depending on the fit of a survival suit, pulling one on can be a formidable task even on solid ground, but it often proves to be an impossible feat while thrashing about in the open sea.

As Breezeman drifted away, Christiansen yelled, "Johnny, how you doing?"

"Help me!" yelled the shuddering Breezeman.

"You hang in there now!"

Christiansen pulled up the all-important zipper on his survival suit and locked it in place. Then he stood and again spotted John Breezeman. The currents had carried him well off now. He was already about three hundred feet in the distance. It was obvious that the man was still struggling to put on his own suit. The skipper realized that he was about to lose one of his crewmen.

"He's not going to make it," Christiansen said aloud as he "bailed into the water after the man." But then, as he swam, he too began to struggle, for his own survival suit was leaking badly.

He swam on, however. It would be a long, hard battle against the wind and currents, with no guaranteed outcome, yet he remained determined to save Breezeman. But the leaden weight of seawater continued to flood into his suit, making it ever more cumbersome, and soon Christiansen's own problems brought heroic rescue efforts to a standstill. His suit was leaking constantly, and he now found himself caught up in a desperate struggle just to remain afloat.

He was working hard to tread water and remain on the surface when he spotted a piece of flotsam drifting past him over the waves. It was a chunk of decking six feet wide and eight feet long. Christiansen decided he'd swim for it. If the piece of decking was buoyant enough to support him, he might still be able to paddle close enough to the struggling John Breezeman to save him.

But as Christiansen swam toward the broken chunk of wooden decking, he became tangled in a drifting maze of crab-pot lines. He fought to free himself.

The *Angela Marie* was barely afloat now. She was drifting along in the current with only the uppermost portion of her bow still showing above the surface of the sea. Charlie Christiansen was certain that the lines holding him were still attached at the other end to her sinking hull. For each time the ship dipped into a wave trough, she would sink from sight and the lines wound around Christiansen's leg would jerk him underwater.

He held his breath each time as he was pulled nearly a fathom beneath the surface. Then as the vertical drifting form of the *Angela Marie* rose over the next wave, Christianson would paddle back to the surface, inhale deep lungfuls of the precious air, and resume the fight to free himself before the next frigid dunking.

Once, twice, three times, and yet a fourth, Christianson was pulled under. He was struck by a terrifying notion: What if the Angela Marie sunk altogether? I've got a wife and three children, he thought, his terror turning to resolve. I've come this far, and there's just no damned way that this boat is going to pull me down now.

Gathering himself, he dived under, reached down, and managed to slide the half hitch off his ensnared leg. When Christiansen surfaced, he spied the wooden chunk of flotsam, now 150 feet in the distance, and immediately began swimming for it. He paddled slowly up the face of each new wave and slid down into the deep valleylike wave troughs. When he drew close enough, he grabbed the nearest edge of the floating decking, dragged his sea-numbed body halfway up onto it, and tried to gather his thoughts. He would rest and catch his breath. Then, as he watched, the bow and mast lights of the *Angela Marie* sank from sight. They never reappeared. The ship was gone. Charlie Christiansen had freed himself at the last possible moment.

Once again, Christiansen spied the drifting, struggling figure of John Breezeman. He appeared to be a full football field or more away. It would be a long journey on such an awkwardly shaped object, but he had to try.

Then in the distance he saw the ship's life raft explode to the surface. It inflated almost within arm's reach of the numbed and astonished Breezeman. Christiansen saw Breezeman reach out, catch hold of the raft, and slowly pull himself aboard.

Later, the wind blew the life raft past Christiansen. It was several hundred feet away at the time, floating up and over the steady onslaught of breakers, and as it passed, Christiansen caught sight of John Breezeman huddled inside, underneath the raft's protective dome. Breezeman offered a simple wave as he floated by. Charlie Christiansen took comfort at the sight. But the raft was too far off and was being pushed much too quickly by the driving winds for Christiansen to give chase. All he could do was to cling to the wooden chunk of flotsam and pray for someone to rescue him and his crew.

As crewmates Kurt Kivisto, Alann Erickson, and Erik Kegal drifted toward shore, they felt hopeful, even exuberant.

"We've got her made," yelled Kivisto. "We're headed right for the beach there!"

They paddled hard but as they neared the shore they discovered there was "No beach at all," just a deadly stretch of coastline. As the three crewmen drifted ever nearer to shore, the waves began to stack up and towering seas swept past them. The crewmen could see huge piles of boulders, some larger than dump trucks, piled precariously on a sharp incline of inhospitable rock bank. Those stretches of waterfront that were not strewn with huge rocks or rock slides were fronted with sheer granite cliffs. The sound of waves colliding headlong into unforgiving rock produced a deafening series of explosions, sending foam and spray fifty feet and more into the frigid Alaskan sky.

Now the boisterous confidence of the crewmen clinging to the cluster of buoys gave way to open fear. As they drifted closer to the cliffs, they gasped at the tremendous flood and ebb of the rampaging surf.

"Hey! Take a look at that," yelled Kurt Kivisto. "Just look at those waves and those cliffs! We don't want to go in here — we won't make it if we do. Let's try to make it around the corner there."

The image of being splattered bodily across the rocks or slammed against the sheer face of the cliff terrified the crewmen. Holding on to the buoys with one hand, the three crewmen began kicking and stroking against the push and pull of the breaking seas. They swam for their very lives. Spurred on by the nearness of death, their adrenaline-filled bodies worked with almost superhuman vigor.

Alann Erickson was almost blind without his glasses. And once, as he paddled on his back, he spied what he thought was a plane and yelled to his crewmates.

Kurt glanced up. "You dumb bastard!" His voice was tinged with a blend of humor and disgust. "That's an eagle!"

Unable to see the cliffs, but paddling mightily, Erickson often asked, "Are we going to make it? Are we going to make it past the cliffs?"

"Keep paddling" was the constant reply.

Then it was Kurt Kivisto's imagination that got the better of him. "Hey, listen!" he yelled. "I think I hear a jet!" But it was only another gust of williwaw wind exploding down out of a mountain pass. As they watched, it tore at the water even as it raced across it — a gray-black cloud of indiscriminate power.

As the young men paddled determinedly on, the wind and tide rip worked against them. The men passed quickly from discomfort, to fatigue, and then to utter exhaustion. But not once did anyone complain or stop paddling, and as they swam, they prayed silently for a plane or boat to come rescue them.

* * *

Meanwhile, Charlie Christiansen found himself drifting alone over the tall gray seas. It was not the storm waves, but the direction the currents were carrying him that worried him sick. For he knew that in such weather, Cape Bingham offered one of the most nightmarish stretches of shoreline a drifting crewman could face.

As he floated in closer, his worst fears became reality, for Christiansen found himself faced with what he described as "a tremendously craggy shoreline—a horrifying stretch of beach." As he drifted into the kelp beds, he could see huge frothing breakers rising up just ahead of him. Pushed by the wild winds, the waves were stacking up as they neared shore. Due to a combination of wind, current, and the rising seafloor, the breakers seemed to take on new life. They gathered unto themselves unforeseen powers, each one mounting up and launching forward on a final frightful headlong rush to shore.

Christiansen could see them exploding against the shoreline, and within himself he rebelled.

Bullshit! he thought. I'm not going in on that piece of shoreline! He wanted to swim around the point to try to search out a safer stretch of shore. But it was a long way, and he was forced to admit that he'd grown too cold and tired to try to buck against the tide rips and wind now driving him into shore.

He flashed to the mistake an old skipper and good friend of his had made when his ship had sunk years before. He and his entire crew, Christiansen recalled, had managed to abandon ship safely. But the skipper had been killed trying to reach shore through the surf. The pounding waves had proven too much for him. The man's name was Wilbur Owen, and the boat was the *Aloha*. It had gone down just outside of Sitka back in the 1970s.

Christiansen wouldn't be drawn to his doom that easily, he decided. He released the wooden chunk of flotsam he'd clung to for so long. Shortly, it was swept away and lost, vanishing into the misty roar of the pounding surf awaiting him just ahead.

His mind raced. How could he keep from being sucked into the surf and lost forever? The desperate, inventive answer came to him suddenly. It was drifting all around him—kelp! Somehow he would tie himself to the floating tentacles of kelp and try to hold on until help arrived—

He turned then, grabbed several long fronds of kelp, and wound them around his arms. He thought of his wife again and his children. He decided he'd hold on forever if he had to. Numbed and tired as he was, Christiansen feared he wouldn't be able to hold on for long. For now, as he waited, the building surf seemed to seek out his stationary figure. He could hear the terrifying roar of the waves breaking well before they reached him. Then he and the kelp fronds began to float up as the steep green face of a wave lifted them. But the kelp extended only so far; when its length played out, the tentacles wound around his arms came tight. Unable to rise higher, Christiansen found himself buried as much as a body length beneath the surface as the gathering surf wave passed over him.

Initially, the call the Coast Guard received was a "classic 'We're going over. Come and help us!' message," recalls the often-decorated U.S. Coast Guard helicopter pilot Capt. Jimmy

Ng. Harried follow-up calls from concerned fishermen in the area enabled the Coast Guard to determine where the crab boat *Angela Marie* might have been when the call for help was made.

Within seconds, Jimmy Ng and his helicopter crew climbed aboard and lifted off from their base in Sitka. He and copilot Lt. Com. Mark Guillory were flying a new $14 million state-of-the-art Black Hawk H-60 helicopter. They cruised at an altitude of five hundred feet and at a speed of 150 knots. As they closed on the general area of the Mayday call, they searched the windswept waters below for signs of life. "We were getting beat up pretty good," recalls Ng. The battering wind forced Ng and his crew to "buckle themselves in," a precaution rarely taken in the ultrastable Black Hawk.

One thousand feet, Ng knew, was about the right altitude for getting an overall picture of things floating in the water. But he also knew it was not the best altitude for seeing an individual crewman drifting on the sea below. The lower altitudes were better for locating survivors.

Captain Ng traced the run of a williwaw gust, perhaps a hundred knots of opaque power, racing down out of a nearby mountain pass. It struck the ocean surface "like an invisible fist," launching an explosion of salt water several hundred feet into the air. Then Ng spotted something brightly colored floating at the base of a craggy cliff-lined stretch of coastline several miles in the distance.

"Hey, I think I see some buoys—or a raft—floating in the surf over there!" Ng announced suddenly to his crew.

Ng banked the chopper sharply and swung in for a closer look. As he approached, a cluster of bright reddish orange buoys came into clear view. But no crew members appeared to be nearby. The buoys were bobbing violently on the edge of a mighty surf that was exploding high against a sheer cliff face.

Ng descended quickly. He swung down over the buoys and hovered there about fifteen feet above them. Not until then did he spot the three struggling crewmen. Two were clinging to outside buoys. They looked healthy enough, but the one drifting on the inside of the buoy cluster appeared "almost unconscious."

"Stand by to deploy!" said Ng to his rescue crew. "And in-flight engineer—get ready to hoist!"

Ng maneuvered the craft in close and swung in low over the ocean swells, hovering just off the water between the cliffs and the raft of buoys and men. The blades of a hovering Huey helicopter produce a sustained ground wash of approximately 150 miles per hour, and by placing his craft in such a position, Ng was able to literally blow the men and buoys up and over the surf waves and back out to sea.

As the chopper hovered overhead, Erickson tried to motion from the water that there were two more men in the water just around the point, but his gestures were lost on the helicopter crew. First things first: The chopper crew kicked out a life-raft canister. Then U.S.C.G. rescue swimmer Dave Beachem, complete with mask, fins, wet suit, and snorkel, leapt feetfirst out the side door of the helicopter, surfaced, swam to the raft, and quickly inflated it. This would provide the waterlogged crewmen from the *Angela Marie* with momentary shelter while the rescue team assessed the situation.

Alann Erickson paddled tiredly over and, aided by the diver, crawled over the side and into the life raft. He, in turn, pulled Kurt Kivisto and then Erik Kegal aboard. When the diver finally came aboard, Erickson rushed to fill him in.

"Hey, there's two more guys up the coastline there," yelled Erickson. "We don't know if they're alive or dead."

Beachem grabbed his waterproof radio and shouted to make himself heard above the roar of the helicopter still hovering overhead. "The guys here are cold, but they seem to be okay for now. They tell me there're apparently two more guys around the point. Better go and take a look!"

Ng hesitated. He'd been gazing at a distant point of craggy shore only moments before when he thought he'd seen what he described as a "minute speck" of something colorful drift up and over a surf wave. Captured in his peripheral vision for but an instant, however, it was far from a clear image. Now, as he held the hovering chopper rigidly in place, he lifted his eyes and prayed that if what he had glimpsed was important, it might show itself once more.

Long seconds passed. Then something orange flashed again. At first, it looked like a life raft with a stick of some kind jutting up into the sky. Then he recognized it as a man in a survival suit with one outstretched arm extended overhead.

Ng accelerated full-throttle toward the distant vision—the form of the *Angela Marie*'s skipper, Charlie Christiansen.

Christiansen was about played out when the U.S. Coast Guard helicopter came roaring around the corner. The chopper banked sharply as it skirted the tops of the cliffs, cutting a magnificent course as it came.

Of Ng, Christiansen would later say, "He was an incredible pilot! He came around that sheer rock cliff going seventy miles per hour and he was right on it. I don't think he and his crew were twenty feet off that rock wall when they were picking me up!"

With Charlie Christiansen now on board, Jimmy Ng lifted off out of the surf, rising skyward to safety. But he had only begun to gain altitude when he spied the *Angela Marie*'s life raft. It had drifted inside the surf and was being pounded mercilessly against the cliffs by the waves. Ng and his crew knew that one crewman was still missing, so he circled in close. The life raft looked empty. They were making a second slow pass harrowingly close to the face of the cliff when, Ng recalls, he and his copilot spied the "classic picture that we love to see in the Coast Guard—a white face, a head, two little eyes, a nose, and a mouth" pushing through the door flaps. The face peered tentatively up at them from under the raft's dome cover. It was crewman John Breezeman, given up as lost by most of his crewmates.

The moment Breezeman was lifted aboard the hovering helicopter, Ng returned to Dave Beachem, his diver, and the three crewmen drifting in among the buoys. Inside the life raft, Beachem explained how they would be lifted aboard the chopper. "Look," he yelled, "the way I'm going to do it is, I'm going to take you out one at a time. Now I know you don't want to go back in the water, but we're going to do it this way anyhow." Beachem would guide each of the three fishermen safely into the body basket. And, one by one, they would be hoisted up to the hovering chopper. He, himself, would go last.

Kurt Kivisto and Erik Kegal were sitting down in the bottom of the raft, shivering uncontrollably, when the rescue basket was lowered to them. Erik went first, then Kurt. Then Alann Erickson swam into the basket. They hoisted his dripping figure high into the air, but when they went to swing him in through the side door, the steel basket became entangled with the helicopter's door handle.

When Alann Erickson was finally pulled inside the hovering helicopter, he found Charlie Christiansen and John

Breezeman waiting for him. Also there, of course, were Kurt Kivisto and Erik Kegal. The entire crew had made it safely on board. For a moment, they smiled at one another. Then their smiles erupted and they screamed and shook hands and hugged one another in jubilant celebration.

But there was a final chapter to their ordeal, even after the crew had escaped their watery hell. In a final, harrowing twist, Jimmy Ng was forced to declare an in-flight emergency when his flight control stick, his "collective," became inexplicably stiff and unresponsive.

In Sitka, the authorities took safety measures in preparation for their imminent arrival by closing the town's only bridge allowing traffic to the airport. They closed the entire Sitka Airport, as well. And when the malfunctioning helicopter arrived, the crewmen could see ambulances, fire trucks, and rescue teams, their lights ablaze as they awaited the helicopter's arrival and possible crash landing.

While the rescued fishermen joked with one another, the Coast Guard flight crew sat stone-faced all the way into Sitka. No longer able to hover, Jimmy Ng and his copilot, Mark Guillory, were forced to land the H-60 chopper like a jet, a maneuver they performed safely. "There must have been two hundred people lined up there along the runway to see us land," recalls Ng. "It wasn't until they got us into the hospital that they informed us that they had closed the bridge down — and the whole airport, for that matter."

ON THE ROCKS

One-hundred-knot winds were howling across Nikolski Bay when, late on the night of March 28, 1987, the 153-foot black-cod catcher–processor *Alaska Star* first struck bottom. Carrying a crew of fifteen, she had come seeking shelter from the vicious Aleutian Island winds on the lee side of volcano-studded Umnak Island. Mike Doland, the navigational master, was at the helm at the time, creeping ahead through the pitch-black night in search of a place to drop anchor.

In addition to using the radar aid that would guide him into the protection of the island, Doland finally decided to switch on the mast lights. He was still moving forward when the slow-warming sodium lamps jumped to life. Suddenly, he was faced with a nightmarish illumination.

"Rocks! There're rocks dead ahead!" yelled one crewman standing in the wheelhouse nearby.

Longtime deckhand Gary Garman was in his stateroom

when he heard the screams. Almost simultaneously, he felt a crunch as the ship ran aground. Garman hurriedly wiped the moisture from the face of the porthole near his bed and peered out into the night. He could make out the craggy shoreline of Umnak Island. It was a bleak setting of "white and black and silver and gray and darkness. And the wind was just smoking!" Slanting streaks of gray-white snow shot across the wilderness setting of sea and shore. He "could have thrown a football" from his stateroom to the rocks.

Then he heard the engines roar and felt the hull shiver as the reverse thrust of the *Alaska Star's* powerful engine and propeller came into play. There was the sound of metal screeching. Someone, Garman realized, was trying to "back away."

As he leapt from his bed, he heard another scream. "Hey, we ran aground! We're on the rocks!"

He stepped out of his stateroom, and as he ran to the wheelhouse, he was forced to dodge the movements of deckhands fleeing down the hallway. There in the wheelhouse, he came upon Mike Doland, the navigation master. He was the man responsible for driving the two-hundred-ton vessel to and from the fishing grounds. And there was Jerry Miller, the ship's fish master, the man responsible for actually catching the black cod. Miller had been in his stateroom, logging in another fine day's catch of black cod, when the ship ran aground.

Then word arrived from the engine room. "We've got a hole in the bottom," reported the engineer. "And the engine room's taking on water! The water's already almost up to the main!"

With the engine room flooding and the ship still headed back out to sea, Miller took over the ship's controls. He knew that the ship was sinking. He also knew that there were no fishing vessels near enough to come to their immediate aid. If the *Alaska Star* went down offshore, its crew of thirteen men and two women would be forced to abandon ship, completely exposed to the pummeling dangers of the open sea.

A foundering vessel Miller knew, could roll over and sink in only seconds. If that happened in a storm as wild as the one they were now encountering, the crew would be trapped inside in the ship's living quarters. Or, with no time to launch the rafts, they would be washed overboard into the deadly waters of the Bering Sea. With the surf pounding against reefs all around them, and full exposure to the one-hundred-knot winds and mountainous waves awaiting them just offshore, the chances of rescuing a drifting crewman in the black, howling, spray-filled vacuum of night would be remote at best.

With that in mind, Jerry Miller grabbed the jog stick and swung the *Alaska Star* 180 degrees to port. He pushed the throttle control full ahead and pointed her back in toward shore. When he first "got her turned around," Miller couldn't see anything. But as he powered ahead, the mast lights soon illuminated the giant and familiar boulders of the shoreline that now lay dead ahead.

The bow of the *Alaska Star* had closed to within only a ship's-length of the shore when she once again struck bottom. She impaled herself hard on a hidden shoal, and for a time, she sat there, rocking uncertainly.

Without warning, the main engine died abruptly. With a final surge of current, the lights aboard the ship brightened, then flickered out altogether. This unpredictable sequence plunged the ship into a world of total darkness; worse, it left those feeling their way about inside the ship suspended in an eerie silence that was terrifying and complete.

With the engines knocked out, the *Alaska Star* pivoted back and forth on the rocks. Those trapped inside could hear the constant groan of hull metal grinding against rock. The high-pitched sound echoed the ongoing destruction through the silence of the interior. And, combined with the incessant moans of the wind howling through the rigging overhead, it sent involuntary shivers racing up the spines of crew members throughout the ship.

"Once we lost power, we started hearing things," recalls crewman Gary Garman. "We could hear the wind and the boat creaking, and we could feel a list growing to starboard."

Though they were perhaps only the ship's own length from the shore to the island itself, in the savage hundred-knot winds, swells from eight to ten feet high drove constantly in against her grounded hull.

After some twenty-four hours of pollack fishing near Seguam Pass, the 125-foot highline trawler *Neakanie* and her exhausted crew were en route back to the protection of an Aleutian Island bay when they picked up a distress signal. The voice of Jerry Miller was controlled and yet deadly serious.

"*Mayday!* Mayday!" he began. "This is the fishing vessel *Alaska Star!* Will anybody who can hear me please come back?"

Frank Bohannon grabbed his mike and rushed to answer the call. "This is the F/V *Neakanie,*" he shouted "Where are you?"

"We've run aground in Nikolski Bay," replied Jerry Miller. "We've torn up our hull, and we're taking on water!"

Without hesitation, skipper Frank Bohannon altered his

heading and began powering ahead toward the foundering ship with all possible speed.

"We're on our way, *Alaska Star,*" replied Bohannon on his powerful marine-band radio. "But it's rough going out here. We've got about three hours of running time separating us. So just be patient if you can. You've got to stay on board and just hold on until you've got someone standing by to pick you up."

Their new course took the *Neakanie* on a long, battering ride through the wave troughs, running broadside to the driving waves.

"It was blowing one hundred knots outside sustained," one crewman aboard the *Neakanie* would later tell me. "We were thirty miles on the lee side of the island and we were taking twenty-foot breakers over the bow."

Regardless, Bohannon maintained their course and speed, powering ahead through the night toward the urgent engagement awaiting them.

Still trapped in the darkness of the *Alaska Star,* Jerry Miller decided to take a look around the ship and check on his men. He knew there were some very frightened crewmen who were already talking about abandoning ship. Without a rescue ship waiting, and with the *Neakanie* still hours away, that kind of strategy was suicidal.

Miller tried to prepare himself, but when he stepped out on deck the wind still staggered him. He made his way along the railing to the back deck. There he discovered that the life rafts had already been launched. Half the crew had already climbed into them and were sitting there, ready to shove off.

"Get out of the life rafts and climb back on the boat!" he screamed above the deafening howl of the wind. "We're not going anywhere! There isn't even a vessel out there to pick us up yet! If you guys take off in those things in this kind of weather, there's a damned good chance no one will *ever* find you!"

Unbeknownst to Miller, back up in the wheelhouse of the *Alaska Star* things were coming apart.

Frank Bohannon was steering the *Neakanie* through "incredible winds and seas that were big and getting bigger," when a frightened voice from the bridge of the *Alaska Star* sounded over the radio.

"We're going! We're going down! We're getting off! We're abandoning ship!" called the voice.

"No! No! Stay on your vessel!" Bohannon insisted, trying to calm the terrified man. "We're only a couple of hours away from you now! You guys stay on board! If you leave now, there's a good chance we won't be able to find you in this stuff!" Jerry Miller caught the gist of the conversation as he entered the wheelhouse, and he took over the microphone.

"*Neakanie! Neakanie!*" he called. "This is Jerry Miller aboard the *Alaska Star.* Yes, sir, we understand that. Our life rafts are in the water, but we'll hold tight here until you arrive."

But the *Neakanie* wouldn't be the first vessel on the scene. For when they had closed to within approximately three miles of the foundering *Alaska Star,* the ship came upon two more ships: the 137-foot *Gunmar* and the 110-foot *Margun,* both owned and operated by the Ilhudso clan, Gunner senior and Gunner junior.

It was Frank Bohannon, however, who organized the rescue and passed on the latest status reports to the nearest U.S. Coast Guard base six hundred miles away in Kodiak. And it was Bohannon who made the decision to maneuver in through the reefs and shoal water to try to reach the crew of the *Alaska Star.* He knew that to run aground and lose a prop or rudder under such circumstances would jeopardize not only his ship but a host of lives, including those of his son, Ethan Bohannon, and of his three crewmen, Rob Frazier, Dale Dickenson, and Craig McKay—not to mention his own. But if they decided to wait for daylight and the *Alaska Star* slipped off the rocks and rolled over, fifteen innocent men and women would almost certainly die.

"At the time," recalls Bohannon, "the attitude on board the *Neakanie* was that we didn't want to lose a single person. Whatever it took to rescue the crew off the *Alaska Star,* that was what we were going to do."

As he pushed ahead then, he found his vessel battered by incredible winds. They tore his 120-knot wind gauge from its mounts and laid the *Neakanie* over on her side.

"The weather was so tough it heeled the boat over at a thirty- to thirty-five-degree list," recalls Bohanon, "and shoved her sideways at a speed of from six to eight knots! The *Neakanie* is powered by a pair of three-seventy-nine diesel engines. They put out about twelve hundred diesel horsepower at twelve hundred rpm. We had to run them at one thousand rpm just to hold our ground. Anything less than that and we'd have lost our steerage."

The seas, too, "were unbelievable," slapping her about as if she were no more than a toy.

With his view of the reef and shoreline hidden by darkness, and with ocean waves breaking hard against the *Neakanie*'s wheelhouse twenty feet above her normal waterline, Bohannon maneuvered stealthily ahead. Unable to see even a boat's length ahead of them, Frank Bohannon and his crew used the radar as their primary navigational aid. At regular intervals, Bohannon called out the distances of three different points visible on the beach ahead and in the bay around them. His son, Ethan, stood at the map table behind him and worked to pinpoint their ever-changing positions.

The typhoon-strength winds thrashed the water into a white lather, camouflaging the possibility of reef or shoal beneath its churning face. With the air trapped in the frothy face of the sea, Bohannon's depth finder was rendered useless. And as he ventured ever nearer, there was no way he could be sure that in the next instant he, too, would not run aground on an unseen shoal or be impaled on an invisible reef or pinnacle of rock.

With reefs on either side, shoal-water markers showing everywhere on their charts, and snow and spray shooting horizontally through the howling black spaces surrounding them, the undaunted crew of the *Neakanie* pushed on.

Every few minutes, the face of their navigational lifeline — the radar screen spinning atop the wheelhouse — froze and their radar system stopped working altogether. Then the crew of the *Neakanie* would tie themselves to one another with rope line and then to the *Neakanie* herself; moving on their hands and knees, they belayed their course up the lee side of the tall steel wheelhouse like mountain climbers scaling a peak. Once on top, a shivering crewman scraped the frozen face and structure of the radar screen clear of ice. Then the crew of the *Neakanie* would gladly retreat, hurrying back inside.

Now, as they moved ahead, the raging winds tore a mast light from its mounts. Then an antenna broke off, snapping suddenly like a frozen branch.

On the third pass into the head of the bay, Frank Bohannon thought he spotted the form of the *Alaska Star* on his radar screen. He powered blindly ahead then, and nearly ran aground "on a reef as long as a football field" before he spotted it.

The foamy black face of the "rock" jutted up out of the surf "no more than fifty feet" off the point of *Neakanie*'s bow.

"We didn't hit, but I'm telling you, we were so close that I had to back down to keep from doing so! Then, of course, I realized what I'd done," Bohannon recalls.

Only at that moment did he see the long invisible form of the *Alaska Star,* and then only because the crew of the vessel had tied tiny, blinking, battery-powered buoy lights along her hand railing.

"We can see you!" came the frantic voice of a crewman aboard the *Alaska Star.* "We can see your lights!"

Regardless, Frank Bohannon knew that, for the present, he could go no farther. He retreated to a point approximately three-quarters of a mile away, where he and his crew would stand by and await the first light of dawn.

Bohannon called the U.S. Coast Guard base in Kodiak with the news: "Am standing by vessel *Alaska Star,* one-hundred-and-fifty-foot long-liner, with fifteen people on board, winds over one hundred miles per hour, gusting now to perhaps one hundred and twenty. Everyone on board here is ready for action should boat roll over."

Bohannon and the other skippers knew that if the *Alaska Star* remained grounded on the rocks as she was, the fifteen crew members on board her had a fighting chance. But if she slipped off the rocks and rolled over during the night, she would spill her crew overboard into the surf. And then it would be a catch-as-catch-can scramble for the survivors. Only the very lucky would survive.

To make matters worse, rescue efforts were hampered, Bohannon later said, by a "bunch of drunken Aleuts," who kept calling ridiculous messages over the radio.

"Get the hell off this channel!" yelled Bohannon finally.

"Aw, man! Don't be like that," answered someone in a drunken slur.

Bohannon and the others anchored up together offshore finally agreed to change radio frequencies. "We had to—to get away from them, and to carry out the rescue," he recalls.

In the wheelhouse of the *Alaska Star,* Jerry Miller was very worried. Seawater had risen up through the subdecks to the main deck, and the stern was sinking steadily.

Once again sensing a growing undercurrent of movement on board, Miller decided to check around. He left the wheelhouse and, bracing himself against the fierce wind, made his way toward the stern of the vessel. There, as he had suspected, he found nearly a dozen crewmen. They had crawled down the *Alaska Star*'s stern ladder and were, once again, sitting in the wave-tossed life rafts.

He could understand their position. The water had long ago filled the engine room and processing deck. Now it had risen to "where it was ready to flood right into their crew quarters." But Miller knew that if they panicked and cast off their lines, they and their life rafts would likely be destroyed on the jagged, wave-lashed reefs awaiting them only a short distance away. And even if their rafts were to somehow miss the deadly reefs, in the wild and impenetrable blackness of such a storm, the crewmen in the crab boats standing by on anchor wouldn't be able to spot them.

The rafts and their terrified occupants would drift out into the icy, 38°F. currents and be lost to the darkness. Far from the close protection of land, they would be blasted by Siberian-born winds of over 110 miles per hour, and pounded by thirty-foot seas. With the freight-train roar of waves collapsing all around them, they would descend into the precipitous black valleys of the giant wave troughs, some stretching two hundred feet from wave top to wave top. Then their life rafts would begin rolling end over end as they are blown down and across the steep, mountainous faces of the waves like so many tumbleweeds; only to be torn apart by the monstrous tonnage of collapsing seas—and lost forever.

"What are you doing?" Miller screamed against the muffling effect of the wind. "Get your asses back on board! Get back on the boat like I told you!"

When the crewmen climbed back on board, Miller confronted them.

"What the hell are you guys doing in the rafts? Didn't you hear me before? Stay inside, out of the wind, where it's dry and warm. I'll let you know when to abandon ship!"

But the navigational master did have a point. One crewman, Gary Garman, refused to leave his life raft. Hey, he thought, there's no future for me on the boat. It's lost. There's no way we can save her. I'm safer here in a life raft and a survival suit than I am on a boat that's already half-flooded with seawater!

The problems continued to mount. While Jerry Miller had been working his way out onto the stern deck to intercept the crewmen climbing down into the rafts, things back up in the wheelhouse were no better.

Frank Bohannon was standing by in the wheelhouse of the *Neakanie* and heard it all.

"God, we're rolling over! We're going! We're going!" yelled the man. "Jesus! Jesus! The boat's sinking! We're getting off! We gotta get off this thing! We're abandoning ship! We're going to the rafts!"

"Wait a second!" retorted Bohannon at the first opportunity. "You guys stay right the hell where you are! Until you're definitely sure that the boat is going to roll over and sink, the safest place for you guys is on board that vessel. If you come out in your rafts now, this wind is going to cartwheel you!"

When Jerry Miller made his way back into the darkened wheelhouse, he grabbed the mike away and, once again, established contact with Bohannon.

"Frank, this is Jerry Miller," he began. "Everything's all right here now. We just had a little misunderstanding. That's all."

Jerry Miller, Bohannon soon realized, dealt with the situation over the radio in a cool and methodical manner.

*   *   *

At daybreak, Bohannon and his crew spotted the *Alaska Star* off in the bleak winter morning. She was listing sharply to starboard. With her stern down and her bow resting just off the beach, she could be seen pivoting back and forth on the rocks that impaled her. Then the weather closed in again and she was lost from sight.

Every now and then in the ensuing hours, the snow flurries and blowing spray would lift, and then, off in the gray nothingness, Bohannan and his crew could again catch a glimpse of the "ghost ship there near the reef."

Bohannon knew there was no way he could get to the sinking vessel. But with fifteen lives at stake, something had to be worked out. Then Gunner senior, on board the *Margun,* came up with an idea. The crew of the *Alaska Star* could fasten a flashing buoy light onto a buoy and, attaching a ground line to the buoy, play all three over the side. Judging by the direction the fierce winds were blowing, there was a solid possibility that Bohannon could maneuver in close enough to snag one end of it. Then, carried along by the wind and seas, the crewmen on board the *Alaska Star* could slide down it to safety.

It was the best idea yet, everyone agreed. But their big break came in the form of a change in the weather. For, along toward daybreak, it had started to ease up—ever so slightly, but enough to make a difference. "I mean, it was still screaming. But you could tell it was starting to back down just a little bit," Bohannon later recalled.

It was still blowing fifty to sixty knots when Bohannon idled nearer to give the crew of the *Alaska Star* something to aim for. Once the plan had been agreed to, the terrified crew aboard the *Alaska Star* worked at a feverish pitch to get the buoy tied off and over the side. Then, fastening a buoy and strobe light on one end, they tossed both over the side, followed by more than half a mile of line. They played the blinking buoy marker and line out in the water and watched as the fierce velocity of the wind and waves swept it away.

Frank Bohannon was idling the *Neakanie* ahead into the gloom when he called to Jerry Miller. "Jerry, the water's too shallow. I'm in as far as I can go, and I still can't see the end of your buoy marker yet. So let out more scope in your line. I just can't get up there any farther. The water's too damned shallow."

Shortly, Bohannon spotted the marker. "Okay, I have it in sight!" he yelled over the radio. "I can see it! But it's still too far up in the shallows. Let out some more line."

Bohannon had his crew drop anchor, and he ordered them to let out more scope (length) in the *Neakanie*'s chain line. Then, gunning her powerful engines, he maneuvered close in an effort to snag the buoy and line.

"Okay! I have it!" Bohannon radioed to the *Alaska Star* a short time later. "And it's tied to my vessel. Go ahead and put your crew in the rafts and send them to me, one raft at a time."

W ith their ship listing sharply to starboard, and water rising up into the wheelhouse itself, the crew of the *Alaska Star* wasted no time in abandoning ship.

"The boat's waiting for us! Let's go! We're out of here!" shouted one deckhand.

One by one, they descended the wave-tossed ladder and slid down inside the dome-covered rafts. Gary Garman would later recall that as they did this, he noticed how the STAR portion of the ship's name stenciled on the stern was disappearing from view, sliding off into the depths.

It was a solemn group that huddled inside the close dark space inside the raft. Gary Garman did not speak again until he and the others were safely aboard the *Neakanie*.

From half a mile off, the faint gray image of the *Alaska Star* was barely visible in the distance. It was snowing hard and gusting to sixty knots when the first raft came drifting out of the blinding maelstrom.

"You couldn't see the vessel too clearly. But as she lay there, you could see a little orange thing detach itself and slide down the lifeline. They pulled themselves hand over hand to us," Bohannon said.

When they arrived alongside the *Neakanie*, Gary Garman felt two crewmen on board the rescue ship grab him and lift him by his shoulders up and over the side.

"You don't know how nice it is to see you," remarked another *Alaska Star* deckhand.

Yet another said simply, "Thanks."

"Hey," said one of Bohannon's crewmen as he helped them aboard, "if you want to kiss the deck, feel free."

Several of the *Alaska Star* crewmen did just that.

Jerry Miller and several other crewmen remaining on board the *Alaska Star* watched as the first raft, carrying more than half of his crew, slid down the rope line and across the water.

"Jerry," called Bohannon, "you and the rest of the guys can go ahead and come over now."

"No, I want to wait," replied Miller. "I want to stay with the vessel. If it gets severe, we'll get aboard the raft and come over."

"Listen, Jerry," argued Bohannon, "there's no reason for you to stay. Get on the raft and come on over. Don't stay there. You can watch it as well from over here as you can from your vessel."

Bohannon was right. Miller turned to the others. "Grab your stuff! We're going!" he commanded.

A short time later, in the gray light of day, the remaining few on board the *Alaska Star* hurried down the rope ladder and lowered themselves onto the second raft. Jerry Miller followed last. He cut them loose from the stern railing. Then he stood in the doorway of the raft and, with everything he had, began pulling them down the line leading toward the safety and warmth of the waiting vessel.

As he pulled, Miller could make out the small, far-off figure of the *Neakanie* rolling sharply in the heavy seas. At the opposite end of the line, he could make out the tiny, haze-shrouded figures of deckhands awaiting their arrival, as well as the inviting glare of the *Neakanie*'s mast lights beckoning him on.

It seemed that they arrived there in just moments. He calculated his boarding attempt to coincide with the crest of the wave and the roll of the boat. Miller flew up and over the side, clearing the railing and landing on deck with a gymnastic effortlessness born of fear and adrenaline.

Then, in the buffetting winds, Frank Bohannon ventured

out on deck. He grabbed Jerry Miller's hand and shook it vigorously. "Jerry," he said with a smile, "I'm glad you came over. You did an excellent job getting everybody off the vessel."

"You're unbelievable," offered Miller matter-of-factly. "We wouldn't even *be* here if it wasn't for you, Frank!"

"The Old Boy upstairs is the one who saved you guys," countered Bohannon. "We were just the ones who happened to be in the right position. He let the wind down just at the time you guys were abandoning ship. If it was still blowing a hundred knots, like it was, I think we would have kissed some of you guys good-bye." Bohannon paused. "Let's go up into the wheelhouse and talk."

Two hours after all fifteen crewmen off the *Alaska Star* had been brought safely aboard, Frank Bohannon, Jerry Miller, and their crews watched the *Alaska Star* slide backward off the rocks and sink, stern-first, into the depths.

"Well, there goes my stateroom," said one as the hull vanished into the depths.

"And there goes mine," called out another.

Still another added, "Well, I hope the wife doesn't mind her picture going down with the ship."

While other crewmen bemoaned the loss of their guns and personal belongings, the *Alaska Star* continued her backward slide. When "she got about three-quarters of the way under, she just slipped, and all but the mast lights and the very tip of the bow" went under with her, recalls Miller.

The *Neakanie*'s crew rushed to secure the deck and accommodate the survivors. Showers were run, dry clothes were exchanged for wet ones, and meals were rustled up for some twenty crewmen and their skippers.

After eating, exhausted crew members off the *Alaska Star* slept wherever they fell. "It was a full house," recalls Ethan Bohannon.

They anchored up in a nearby bay to await the arrival of a bush plane to take the crew of the *Alaska Star* back to town. But tremendous winds still prevailed, and the plane never arrived. Finally, Bohannon boated the entire lot into Dutch Harbor himself.

With the ordeal now over, the crew of the *Neakanie* found themselves completely played out. "My gang," Frank Bohannon told me, "put out one hundred percent. In fact, they put a hundred and fifty percent into the rescue. And afterward, everybody was just washed out and shot."

As Ethan Bohannon recalls it: "When it was over I was weak, tired, and looking for my bunk."

"The guy that saved their lives was a guy by the name of Jerry Miller," Frank Bohannon told me later. "If he'd let them take the boat off the beach, with the son of a bitch sinking like it was, they would have been deader than doornails. Jerry had the presence of mind to assess the situation and to take her and wreck her again, putting her right back on that reef. For me, it was fun. I mean, there was a lot of challenge to it—to use everything I'd learned in twenty-seven years of fishing and see if I couldn't pull it off. It sure got the adrenaline pumping."

The original name of the *Alaska Star* was the *Aku Maru* #33. The words, in their original Japanese, mean "eternal circle."

Some eleven years before, the *Eternal Circle* had been seized by the Coast Guard while fishing illegally inside the two-hundred-mile limit. Later, she had been auctioned at a U.S. marshal's sale. Over the next decade, the vessel was sold and resold; she became a catcher-processor, then a floating processor, and then a catcher-processor again.

When the *Aku Maru* sank to the bottom of Nikolski Bay during that wintry Alaskan storm in 1987, she had been fishing in the same Aleutian Island waters where she'd been seized some eleven years before — thus completing a circle of service, as well as fulfilling the prophecy of her christened name.

THE FACE OF AN ANGEL

*"You blink your eye, and your heart beats twice, and the wind
has come up to be just ferocious, and we're in the middle
of this big goddamn tide rip; and I mean, ten-foot waves are
standing up with their tops breaking. . . ."*
—Mark Hutton

The village of Ketchikan clings romantically to the steep
green-forested hillsides of southeast Alaska. The trim Boston-
style dwellings of old still stand overlooking the surrounding
islands and the bustling salmon-filled waters of Tongass Narrows; far overhead, snow-capped Deer Mountain rises out of
a lush rain forest that has absorbed over two hundred inches
of rainfall in a single year.

Bald eagles are plentiful in this archipelago region of
islands and inland sea. Their gleaming white head feathers
flash brilliantly as they cruise past or sit perched atop the dark
green canopy of an ancient spruce tree.

Considered Alaska's gateway city, Ketchikan is a robust
center of frontier commerce and can be reached only by boat
or airplane. Bears are often chased from within her city limits.

During the summer months, along the waterfront, scores
of pontoon bush planes can be seen taxiing about, docking

here and dodging salmon boats and tour ships there. And throughout the day, scores of these planes roar off across Tongass Narrows, all the while fighting to break their pontoons free of the suction of the water. Finally lifting off, they rise into smog-free skies, carrying people and cargo to one of scores of logging camps and native fishing villages in the wilderness outback. These bush planes raise such a deafening racket upon takeoff that casual conversation along the waterfront is often impossible.

Through the years, in the downtown area of this beautiful rustic village by the sea, the town elders cleverly laid claim to land where there wasn't any. They drove wooden pilings down through the seawater and tideland mud and, striking bedrock, built hotels, whorehouses, canneries, and saloons upon them. The red-light district was closed in the 1950s, but the hotels, trinket shops, and fish-packing enterprises, as well as the wild all-night saloons, still remain.

It was September 9, 1989, when Mel Wisener, skipper of the fifty-foot salmon seiner *Rebecca,* called and asked Mark Hutton to come to work. Wisener was fishing out of Ketchikan at the time and was just finishing up the salmon season there.

"Mark, my crew just bailed out on me!" Wisener began. "Why don't you come on down here for the last couple of weeks of the dog-salmon season? And bring your shotgun. We'll go deer hunting; we'll set out some king and Dungeness crab pots and have a good time."

Wisener also asked Hutton to bring along his longtime fishing partner and best friend, Tim White. For Hutton, the experience sounded like a nice way to close out what had been a long and difficult year of hard work. He agreed to call Tim White on his ranch in Montana.

And so it was early in the fall of 1989 that deckhands Mark Hutton and Tim White descended upon Ketchikan. They would fish a couple of short salmon openings, close out the season with Wisener, draw their pay, and fly home.

The two men met aboard the old wooden boat *Rebecca* and, in anticipation of the salmon opening that was close at hand, tossed their belongings aboard. Then, moving through the quiet peace of early-morning light, they stood out on deck as their skipper guided the *Rebecca* along Ketchikan's scenic waterfront and out of town by way of Tongass Narrows.

Standing six five and weighing 260 pounds, Mark Hutton was a skipper's dream. Hutton had lived in Cooper Landing, on Alaska's Kenai Peninsula, since 1971, and he was a seasoned crewman. He'd been a starting tackle on the Oregon State University football team and, as such, was there when the OSU defense held USC running back O. J. Simpson to a mere three yards of total rushing for the entire football game — a contest OSU won by a single field goal, the game's only score.

Now, after some eight months of fishing, Mark Hutton was tired — "soul-tired" — from it all. He'd started out in February in Petersburg with Allan Ottness on the F/V *Commander,* going out for halibut using long-line gear. Then he fished for herring and black cod and, finally, halibut out of Sitka. He'd flown eight hundred miles north then to Norton Sound to fish another herring opening with a different skipper, then returned to Sitka for yet another halibut season. Finally, he had fished the gillnet season up in the legendary waters of Bristol Bay.

Several hours after their departure from Ketchikan, Mark Hutton was standing in the wheelhouse, "bullshitting with the skipper," when the boat started acting strangely. The two men glanced instinctively out onto the back deck. The starboard side of the deck was already underwater.

"Jesus," said Mel. "Maybe we've got a problem with the way the seine net is positioned out there. Let's get out there and swing the boom and power block into a new position."

When they had completed that task, Hutton and Wisener returned to the wheelhouse. The shifting of the weight of the cargo and block soon brought the back deck up level again, and as they watched, the wave wash cleared from the deck. But only minutes later the deck once again began to list, this time to the other side. And just as suddenly, the port side began to disappear underwater.

At the same time, the *Rebecca* and her crew were moving into "heavier and heavier seas." In an area of converging sea channels, an abrupt force of wind struck them. In the buffeting winds and waves, the *Rebecca* began to respond sluggishly, rolling from side to side with an alarming tentativeness. And once again, the vessel began listing about fifteen degrees.

In the wheelhouse of the *Rebecca*, Mark Hutton turned to his skipper. "Mel, this is your boat! What do we do?"

"Well," replied Wisener, "I think if we just move the vaining winch [the tall steel boom arm] over a little more, we'll shift the weight and bring the deck back up for good."

Respectful of the sea, and conscientious by nature, Hutton questioned that decision. "Mel, I don't like this. I think we should go back to the dock. We could turn around right here and go back to the dock and find out what's wrong."

Mel disagreed. "No, I don't think it's a real problem. I think we can keep going and whatever is wrong will correct itself."

"Mel Wisener wasn't a big guy," recalls Hutton, "but he was as tough as an eightpenny nail." He was the most experienced of all the men on board, and since Hutton respected Mel as both a fisherman and a friend, he yielded to his authority. It was a decision he would soon regret.

As the *Rebecca* began to move out into the tidal rips and windy reaches of Sumner Strait, she listed yet again, this time twenty-five degrees to starboard. As they raced to right her, they were met by a sudden series of steep breakers that, one after the other, began driving headlong into them.

Up to that time, Tim White and another crew member were both still lying in their bunks, unaware of the growing predicament. They "didn't have a clue," recalls Hutton.

"Damn, Mel," said Mark Hutton to his skipper. "We've got a real problem here. I'm going to get the boys up! What are you planning to do?"

"I'm going to go cut the skiff and the seine net off the back deck!" announced the skipper.

Aw hell! thought Hutton to himself. The long and profitable year of fishing had passed without serious trouble. But suddenly, all that was changing now. This can't be happening to me now, he thought. Not now!

"Tim!" he shouted to his crewmate and best friend. "Get your ass out of bed and get dressed! I think we've got a real serious problem here!"

Tim had to go the bathroom, regardless of the drama unfolding around him, and so he stumbled across the slanting floor and stepped into the cramped confines of the head.

Tim White had been raised in the Finger Lakes region of upstate New York. He was still in his twenties when he made the adventurous move to Alaska, settling near his longtime buddy Mark Hutton in the beautiful Cooper Landing area of the Kenai Peninsula. New to Alaska, White was nevertheless quite experienced in the out-of-doors life, and he adapted quickly to river guiding. He had often guided parties of from six to eight people for four days and nights at a time on wilderness float trips in the Twin Lakes area on the Kenai Peninsula. But after a decade or more of guiding, White found he was "getting tired of living in slickers and rubber boots all the time."

Only recently, after some dozen years in Alaska, Tim White moved to Montana to get away from Alaska's dark, cold winters. He moved down into Montana's beautiful Bitteroot Range, where he and his wife bought a ranch, complete with "horses, chickens, ducks, goats, and turkeys."

Now White found himself caught up in the kind of high-seas adventure that one could only pray never happens. As he stumbled toward the head, he took in the odd angle of the walls and the steep list in the floor.

Inside the head, Tim White found himself "practically standing on the wall," and as his brain came awake, he said to himself, Well, damn! This just isn't right! By the time he was ready to leave White knew they were in trouble. He exited the bathroom mad, wide-eyed and charging.

He hit the galley floor on the run and had accelerated no more than two steps when he ran headfirst into the solid lineman's bulk of his good-natured friend Mark Hutton. "You guys better get up and get dressed," Hutton told him.

"I guess!" retorted Tim White, the tone of incredulity rising in his voice.

Then Hutton turned to the other deckhand, who had also appeared in the galley. "Get the survival suits," he ordered him, "and take them up onto the bridge."

There was only one inexperienced deckhand on board, and as luck would have it on that particular voyage, he had left a two-gallon jug of Wesson oil sitting on the kitchen counter. In the rolling seas, the jug had toppled over, spilling its oily, volatile contents onto the red-hot surface of the galley stove.

Tim White knew that such an incident could easily set the entire ship on fire. Smoke and fumes were already coming from the stovetop when White reached the galley. He grabbed an armload of paper towels and did his best to wipe the surface clean, soaking up all the oil he could reach, all the while cursing the situation angrily.

Task completed, White grabbed his survival suit and stepped into it, shoes, pants, and all. He wrapped the long foamy arms of the suit around his waist and tied them in a knot. Then he pried up a chunk of the floorboard and peered down inside the engine room to see if any flooding had taken place. To his surprise, the compartment was completely dry. He rose quickly to his feet then and glanced through the rear window out onto the back deck. He was startled by what he saw. Foamy ocean waves were crashing across the full length and width of the *Rebecca*'s back deck.

Then, turning, he took in the strange disfigured shape of the head door. The face of the door was ballooning out, as if some mysterious force were about to explode through it. White soon realized that it was the force of a rising flood of seawater. The vessel had been listing so sharply that the water had started running backward through the vessel's plumbing. With no check valves to stop the flow, seawater was pouring out of both the head's sink and toilet. It now stood chest-deep in the tiny void of the ship's head.

Shit, we're in real trouble here! he thought.

"Where's Mel?" Hutton yelled to Tim White as they raced up into the wheelhouse.

"I don't know," Tim shouted.

Once in the wheelhouse, White turned and glanced aft

through the spray-blurred glass of the wheelhouse windows. He quickly spotted his skipper, Mel Wisener, standing with his legs braced wide on the steep and pitching deck, fighting against time to reposition the tonnage of the seine skiff and the cumbersome head-high stack of black seine net piled on deck.

White grabbed the marine-band radio's microphone and began calling for help. "Mayday! Mayday! Mayday! This is the fishing vessel *Rebecca!* We're adrift off El Camino Point and we're in trouble here! Mayday!"

Just then, a large wave broke over the ship, rocking her severely. The force of the wave threatened to topple the *Rebecca.* Tim White and Mark Hutton ducked down and held on. The explosive mountain of sea found Mel Wisener still on deck. The wave engulfed him, the ship's seine net, and her twenty-foot skiff as well, sweeping them across the deck and over the side.

Tim White rose and, when he had regained his balance, once again began calling for help. "Mayday! Mayday! This is the fishing vessel *Rebecca!* We are in trouble!"

Then, with microphone still in hand, he shot a glance toward the back deck to check on Mel Wisener. He couldn't believe what he saw. The skiff and the tall black pile of seine net were gone, and so was Mel Wisener.

"He must had gotten washed overboard!" yelled White. "Jesus!"

Tossing the microphone to Hutton, he said, "Send out another Mayday call! Then go up top and break out the life raft! I'm going after Mel!"

Tim White zipped his suit the rest of the way up and headed down out of the wheelhouse. The wave had knocked the *Rebecca* almost completely onto her starboard side. White climbed up the sloping deck and, balancing himself atop the ship's port-side handrail — now lying horizontally — made his way toward the stern of the heaving vessel.

White walked along the uppermost edge of the teetering handrail like an experienced acrobat. He paused at the end of the ship's stern in an effort to locate Mel Wisener before taking the plunge overboard. Then, as one of the ten- or twelve-foot waves lifted the stern, he spied what he believed was the drifting figure of his skipper. He appeared to be well off in the distance, at least a thousand feet away.

White knew that his skipper had been washed overboard without his survival suit, and that, cast adrift in such "nasty stuff," Wisener wouldn't be able to hold out against the cold, brutal elements forever. Getting the life raft inflated was also absolutely essential, and he knew he could depend upon Mark Hutton to do that. When he looked back, he spotted Hutton standing on the almost-flat surface of the outer wall of the wheelhouse, already bent to the task of breaking open the life raft's canister. With that image secure in mind, Tim White leapt overboard.

The cold seawater stung the bare flesh of his face. And when he bobbed to the surface, he immediately turned onto his back and began backstroking out to sea. Now and again, White turned in an effort to get his bearings and locate the drifting figure of his skipper, but each time, he could see only an endless series of steep, collapsing waves rolling over a lonely horizon of silvery sea and sky.

As he paddled away from her, White saw the *Rebecca* sink beneath the waves. He watched her right herself then and float back to the surface, where she lay wallowing.

White had never liked the temperamental feel of the *Rebecca*. But now she even refused to *sink* properly. He was tempted to swim back to the foundering vessel and "either burn her or sink her myself," he would recall later.

He paddled on, however and when he had traveled what he estimated to be a thousand or more feet from the sinking ship, he began calling aloud for his skipper. "Mel! Mel! Hey, Mel!" he yelled again and again. But he heard nothing. He drifted and waited, thinking, Damn, I've lost him. He's gone under!

When the *Rebecca* had rolled onto her starboard side, waves began crashing into the vessel's cabin and engine room through the outer air vents. At the time, Mark Hutton had climbed down through the galley and out onto the back deck, where he found himself walking across the face of what had once been the port-side wall of the wheelhouse.

The white canister holding the life raft was shaped like a forty-gallon soda can. It was mounted atop the wheelhouse, strapped in place with steel bands. Hutton punched the hydro-static release and the canister broke free of its mount; he dropped it into the water, then watched helplessly as it became entangled in the rigging cables.

Hutton climbed farther out onto the now-vertical face of what had once been the top of the wheelhouse and clung to the rigging as he worked to free the canister. As his hands flashed to the task, a nasty series of ten-foot lumpers drove into the *Rebecca* and over him.

The sinking vessel lay there on her side for a time; her back deck rose up in the form of a ten-foot-high wall, standing up in almost a ninety-degree position. The waves alternately lifted and crashed over the foundering length of her hull as Hutton scrambled to adapt and survive.

As he watched, the drifting canister was being tossed against the steel masts and cable rigging. Hutton grabbed the lanyard leading from inside the canister back to the top of the wheelhouse and began furiously pulling out the line, yard after yard. Perhaps a hundred feet of the nylon cord fell into the water at his feet before he reached the end and the line came taut. Hutton gave the tiny cord one final powerful yank. He heard the CO_2 cartridge hiss to life. Then the steel bands holding the canister together gave way, parting suddenly under the pressure of the inflating life raft, and the canister exploded open.

"It worked just perfect," Hutton recalls—except that, like the canister that once held it, the life raft itself now became snagged in the mast poles and rigging cables. If the *Rebecca* rolled all the way over, Hutton knew that it would tear the life raft to shreds. Then he and the others would be cast adrift miles from land in the icy seas; and some of them, he was quite sure, would perish. "It was so scary that my brain just kind of shut off," he recalls.

There was still no sign of skipper Mel Wisener or Tim White. Sickened by the thought of losing his close friends, Mark Hutton contemplated climbing back into the sharply tilting structure of the wheelhouse in an effort to send out more Mayday calls.

But it was too late, for as Hutton worked to free the now-inflated life raft, the *Rebecca* rolled farther and then sank out from under him. The life raft was washed free of the entanglement and began drifting away.

"Go for it!" shouted Hutton to the last deckhand on board.

As the vessel rolled, the deckhand dived off the side of the wheelhouse, swam to the raft, and crawled aboard.

Wearing boots, jeans, and a sweatshirt inside his survival suit, Hutton felt he was pretty well dressed to face the predicament at hand. But then as he dived forward toward the life raft and open water, he felt his feet slip, then immediately snag and become entangled on a submerged cable. It was then that he felt the first inkling of real panic stir within him.

As he struggled to free himself, Hutton could feel the chilling cold of the icy seawater pouring in through the neck area of his survival suit. The sudden shock of the chilly water stole his breath away and quickly sent him into a panic struggle.

As the waves pounded in over the foundering vessel, Mark Hutton remained ensnared in the rigging. Trapped upside down, his feet still entangled in the mast cables, he fought to free himself as his suit flooded completely. Unable to lug the hundreds of pounds of excess seawater now ballooning his suit, he was soon exhausted. Finally, though, Hutton managed to fight free of the entanglement and pull himself upright.

He spotted the life raft and immediately struck out for it. But the raft was being blown away much faster than he could swim, and as he floundered on toward the distant and fading object he swam across a line trailing behind the life raft. He seized it, gratefully, and immediately began pulling himself up and over the waves. His mind raced as he worked to gain on the raft. I don't need this, he thought over and over. This absolutely cannot be happening!

When Hutton reached the life raft, the outstretched arms of the deckhand helped pull his water-bloated figure aboard. Mel Wisener was gone, and Tim White was missing, as well. Mark Hutton lay sprawled across the bottom of the raft, too exhausted to move.

For a time, the two men inside the life raft remained silent, drifting numbly up, over, and sometimes through the steep, close-cropped seas—seas that often broke over the raft and threatened to topple them. And as they drifted, the deafening howl of the wind roared over the bonnet of the life raft "like a jet engine."

As the winds, waves, and tidal rip lifted and rocked them, Hutton caught brief glimpses of the *Rebecca* as her wallowing form drifted ever farther away.

Numbed from the exhaustive search and the interminable swim, Tim White was finally forced to abandon the hunt for Mel Wisener. Now he faced the long journey back to the *Rebecca*. But the ship was drifting steadily away, and the gap between him and the sinking vessel was growing wider by the minute.

It was not until he neared the waterlogged shell of the *Rebecca* that he spied the orange dome of the life raft as it drifted over a wave crest well off from him. While the current was pulling White in one direction, the wind was blowing the life raft in yet another.

Meanwhile, Mark Hutton was worried sick about the missing status of Mel Wisener and his good friend Tim White. He was still lying back, trying to recover from his traumatic experience, when Tim White's head popped in through the side opening of the life raft's bonnet. Hutton was openly thrilled to see him. "Tim!" he yelled as he rushed to help pull him aboard.

Once inside, Tim White lay back and tried to recover from his long swim. He was deeply disappointed. He had hoped that Mel Wisener had somehow found his own way back to the life raft. The thought of his skipper "still out there," drifting alone and clad in nothing but a pair of cotton pants and a T-shirt, tore at him.

Moments later, White pulled back the door flap at one end of the bonnet and peered outside. Suddenly, he turned and yelled to his crewmates back inside.

"Shut up!" shouted White.

"What do you mean?" shot back Hutton.

"Shut up!" White spat again. "I think I heard something!"

White had hardly regained his breath when he turned to Hutton.

"I'm going back after Mel!" he announced.

"And then, without a good-bye or nothing," recalls Hutton, "White dived headfirst out the end of the raft and swam off in an effort to 'follow the sound.'"

White swam toward the area where he remembered last seeing his underclad skipper. He tried to calculate where his skipper might have drifted; he took in the direction of the winds and the angle of the lumbering seas and used them as mental coordinates by which to navigate. By his estimate, he had to swim on a course sharply quartering the oncoming seas if he was to locate his lost skipper.

Back aboard the raft, Mark Hutton could not stand the situation. Damn you, Tim, he thought to himself. I don't want to go back into the water, but I'm not going to let you go back in by yourself. And with that, Hutton, too, dived out of the raft. It was a decision that would very nearly cost Mark Hutton his life.

Moments later, Hutton lost sight of Tim White. Then in his struggle to remain afloat, he lost track of the life raft, as well. And then he began to sink. He had been able to zip up his survival suit only to about the level of his navel. Now he could feel the bloated weight of the flooding suit pulling at him, as well as the increasing pressure of the ocean water pressing in as he was driven down into the depths by the collapsing seas. He knew his strength was draining quickly away, failing him almost entirely now.

"In only seconds, it seemed, the waves were breaking over the top of me. I'd get buried and couldn't get my breath. All I could do was struggle up like my life depended on it. And then I'd get buried by a wave, and then I'd struggle up to where I could get just one more breath. I'd no sooner get that than another wave would break down right on top of me and drive me under, and I'd fight like a dog to get back up to the top and grab just one more breath.

"Then I thought my heart was going to explode. I didn't think I could take another breath. And I thought, This is the end!"

Five more minutes passed, and still Hutton's death struggle continued. He knew he had to remain within striking distance of the surface and oxygen and life. Another ten whole minutes passed and still Hutton fought back.

When Tim White had completed the long swim back to the area of ocean where he believed his skipper was last seen, he began screaming for the missing man.

"Mel! Mel! Where are you, Mel?" he called out over and over. But he heard nothing. Scared and frustrated, White could feel his adrenaline pumping. He began to get angry, then furious.

"Mel! Mel! Goddamn it! *Answer me, you SOB!*"

"It was a lousy situation," recalls White. He definitely didn't want to be there. Then, as he drifted up and over one wind-swept wave, he thought he heard something—faintly. Was it a distant reply, or just another blustery gust of wind whistling past the hood material of his survival suit?

Tim screamed again. And this time, he recognized the comeback. It was the voice of Mel Wisener. Instantly, White screamed a frantic response.

"Mel! Mel! I can't see you yet! Keep yelling! Steer me toward you! Tell me which way to go! Keep yelling!"

The sound of Wisener's voice arrived in incoherent bursts. Tim White did his best to return the call, but often his words seemed to be instantly lost as the deafening roar of the break-ing seas effectively muted him.

Minutes of futile searching later, he spied the frigid blue figure of his "leather-tough skipper" drifting over the crest of a nearby wave. Wisener was sitting on the submerged form of a five-gallon bucket. The bucket was a chunk of flotsam that had washed off the back deck of the sinking *Rebecca,* and Wisener had snagged it as it drifted past. By tipping it upside down and trapping air beneath it, he'd clung tightly to it, and, using it as a buoy, had been able to keep himself afloat.

Buffeted by ever-increasing seas, Mel was now doing his best to remain atop it. He was clutching the bucket's handle, riding it up and over the crest of each new wave like a rodeo cowboy.

Tim White swam close and wrapped both of his legs around the shaking torso of his skipper. "Just hang on, Mel," he yelled above the howl of the wind. "I'll paddle on my back and you just steer me in the right direction, 'cause I can't see where I'm going! Try to spot the life raft. Just keep your eye on the raft and tell me which way to go!"

Then White set out on the long journey back. The instant Wisener discarded his bucket, he grabbed a fold of the crotch fabric of Tim White's survival suit. Acting out of his silent fear, he squeezed that sensitive area with the strength of a death grip.

"Goddamn it, Mel! Let go!" Tim White yelled repeatedly as he paddled along. En route back to the area where he had last seen the life raft, White repeatedly asked Wisener about his condition. "Are you all right?" he demanded.

His thoroughly chilled skipper did not reply. Instead, he began yelling, "Mark! Mark! Hey, Mark!"

Tim White's mind raced. Damn, he thought. Now Mel's hallucinating! And what if he dies? I can't just turn him loose.

"Mel! Are you okay?" White asked again.

But Mel Wisener only kept yelling, "Mark! Can you hear me? Mark! Over here!"

Tim White pivoted around in the water then and took a look for himself. To his astonishment, there indeed was the body of Mark Hutton drifting vertically through the water just below the surface. With his hood down and his bloated survival suit unzipped to the waist, Hutton was clearly drowning.

Hearing nothing of his skipper's calls, and using the last of his strength, Mark Hutton fought his way up to the surface and inhaled another lungful of precious oxygen. It would be his last, he was sure. The mental stress involved in such a terrifying experience now seemed to alter time itself. And the seconds began to stretch into a kind of eternal frozen present.

Submerged and slowly sinking now, Hutton's mind raced. The time had come when he could no longer reach the surface. He thought of his funeral. He contemplated his beautiful wife, Martha, with whom he had never been able to spend enough time. This is a lousy way to go, he thought. And then he felt a surge of contempt and anger jolt through him—anger at himself for not being able to fight the battle any longer, contempt for being caught in such an impossible predicament.

He was completely played out, and he knew it. He couldn't even raise his arms to attempt to swim toward the surface. He was a goner. It was over—the end.

"God, I need some help," prayed Hutton ferverently. "And I need it now!"

Mark Hutton had started to inhale seawater when he felt a forceful hand grab him by the hood of his survival suit and yank him toward the surface. The force lifting him felt almost angelic—as though a power unlike that which he would "attribute to a human being" had intervened to save him.

When his face broke through the surface into the cold, fresh Alaskan wind, he inhaled spasmodically and gasped for air. Once he could see again, he found himself staring into the face of his crewmate Tim White. And it was definitely *not* the face of an angel.

"Hutton," he said in mock disgust. "You pussy!"

Actually, Tim White's first vision of Hutton had startled him. Seawater was draining from Hutton's mouth and both his nostrils, and only the whites of his eyes were showing. White grabbed Hutton and held on to him for a few moments to allow him time to breathe and to pull himself together. Then he shouted into Hutton's face, "Goddamn it, Mark. I need your help! So get your act together right *now!*"

Minutes passed, and the three fishermen grappled with the problem of remaining afloat and alive. Tim White was holding on to Mark Hutton; Mel Wisener was holding Tim White, his hands still frozen to the crotch material of White's suit; and while Tim White wrestled to free himself of Mel Wisener's grip, he struggled to keep all three of them afloat and together and to locate the orange dome of the life raft that was now nowhere to be seen.

As they drifted up and over a black and monotonous succession of steep swells, Tim White caught a fleeting glimpse of what he believed was a salmon boat on the horizon. Unbeknownst to him, it was the wooden seiner *Marysville.* Her captain and crew had heard the Mayday call and had immediately begun to search.

She was crashing through the storm waves with an almost hysterical abandon, and as she fought her way toward them, White took in the hopeful flash of exploding seas as the vessel leapt over the sharp swells and disappeared into the deep wave troughs below.

The crew of the *Marysville* wasted little time in spotting the drifting crewmen, and her skipper maneuvered in close to attempt the rescue. A line had to be tossed to the tired and foundering crewmen. Then, one by one, they would be pulled back to the *Marysville.*

But there was a problem: The deckhand on board the wooden seiner who had been assigned the task of throwing the life buoy was so shaken that each time he attempted to toss it to the drowning crewmen—now only a few but critical yards away—he would accidentally step on the rope line attached to it. And each time, the life buoy would come to the sudden end of its short, feverish run and recoil back aboard the rescue ship.

Each new effort by the clearly frustrated crewman brought with it another failure. The angry, disbelieving voice of Tim White began to rise. He began roasting the young man for his ineptitude. "If you don't get that buoy out here to us, I'm going to come up there and I'm gonna break both your arms and both your legs. And *then* I'm going to hurt you!"

But again the young crewman failed to reach them with the hook.

The skipper of the old wooden seiner was steering into the oncoming wind and waves in an effort to hold a forward position, all the while edging sideways ever closer to the drifting crewmen.

Each time the *Marysville* rolled, Tim White could see the vessel's prop spinning wildly, churning up air and chopping through the ocean surface. White knew he couldn't maneuver with his two crewmates in tow. And he worried about the deadly blades of the ship's prop—that and the threat of being crushed by the rolling tonnage of the forty-six-foot *Marysville*.

Half a dozen panicked attempts later, the grappling hook and trailing line finally landed close enough for Tim White to grab. He quickly pulled the attached life buoy to him. He loaded Mel Wisener into the ring and signaled those on board the seiner to pull, and they quickly hoisted him aboard.

With his legs wrapped around Mark Hutton now, Tim White drifted patiently over the lapping seas and kept a close watch. He wanted to make certain that Mel Wisener went first. Once Wisener was aboard, he hollered to the crew of the *Marysville*. "Get his clothes off and get him in a bunk! You'll need to do something, because he's been in the water for a long time!"

The anxious crew quickly stripped Wisener of his clothes, rubbed him down with dry towels, wrapped him in wool blankets, stuffed him inside the warm folds of a sleeping bag, and tossed him into an empty bunk.

Then White turned to Mark Hutton. "It's going to take everybody to drag your sorry ass on board!" he said as he slid the life buoy around him. "So you wait here and let me get aboard first so I can help."

With Mark Hutton safely tethered to the life buoy, Tim White swam to within arm's reach of the side of the pitching, rolling seiner. And when the vessel's side railing dipped momentarily underwater, he slung both of his arms across it and held on. When the boat rolled sharply back and away again, the railing lifted him clear out of the sea and he was yanked instantly aboard.

"The big guy's survival suit is full of water!" he told the crew of the *Marysville* as he stood on her back deck. "You might need a picking boom to lift him out!"

Still adrift in the icy Gulf of Alaska seas, Hutton had all but collapsed from the frigid ordeal. He could no longer "lift so much as a finger," and it took the supreme efforts of six deckhands to pull his more than four hundred pounds of body and water-bloated survival suit up and over the side of the vessel. Hutton crawled to the hatch cover in the center of the boat and collapsed on his back, once again too exhausted to budge.

Then as if out of nowhere, a U.S. Coast Guard SAR boat came pounding through the heavy seas and pulled up alongside the wooden seiner to help transfer the three survivors. Clearly hypothermic, Hutton and White were quickly stripped of every stitch of body covering and wrapped inside the thick, snug folds of multiple layers of woolen blankets. They were safely aboard. They had made it. And then they cried.

That night, another seiner boat, miles away, was making its way through choppy seas when it came upon the drifting form of the *Rebecca*'s life raft. The *Rebecca*'s young deckhand was found aboard. Though he was rescued and returned to Ketchikan, Hutton, White, and Wisener never saw him again.

That night, the three tired but grateful crewmen were released from the hospital. After first buying some clothes to wear, they decided to celebrate their survival by spending a night out in that wild and scenic fishing port of Ketchikan. They had experienced more fear, joy, and adventure in that single accelerated day than many people do in a lifetime. And now they set out to "hit every bar in town."

Their night of frolic and fellowship concluded down on the waterfront. It was there that one could have found them in the final, fanatical throes of inebriation, crawling up main street on their hands and knees. As they moved slowly along, these fishermen, who had cheated death and experienced the bonding that only survivors can know, paused now and again to lift their heads and howl at the gracious heavens.